books by the author:

Behavior
Pediatrics
ive Development of Brain and Behavior in the Dog
avior of Wolves, Dogs and Related Canids
anding Your Cat
anding Your Dog
s in Ethology: Animal and Human Behavior
Animal and Man, The Key to the Kingdom
d Canids: Their Systematics, Behavioral Ecology and
tion

THE
DOG
Its Domestication and Behavior

THE
DOG

Its Domestication and Behavior

Michael W. Fox

Director, Institute for the Study of Animal Problems
Humane Society of the United States, Washington, D.C.

 Garland STPM Press
New York & London

14199

15   14   13   12   11   10   9   8   7   6   5   4   3   2

Library of Congress Cataloging in Publication Data

Fox, Michael W   1937-
  The dog.

  Bibliography: p.
  Includes index.
  1. Dogs—Behavior. 2. Canidae—Behavior.
3. Domestication. 4. Mammals—Behavior. I. Title.
SF433.F69   599'.74442'045   76-57852
ISBN 0-8240-9858-7

Garland STPM Press.
A division of Garland Publishing, Inc.

Printed in the United States of America

Contents

Acknowledgments

The author gratefully acknowledges the collaboration of the following undergraduate assistants and colleagues in the completion of some of the material included in this book: Keith Kretchmer (Chapter I), Ellen Blackman and Dr. Alan Beck (Chapter 3), and James Cohen (Chapter 4). I am especially glad for the opportunity to include some of my Army research data collected in collaboration with Dr. Jeff Linn (Chapter 10) and to reevaluate this material from the original report (Linn, 1974) in relation to other data derived from wild canids and their hybrids in my laboratory. Thanks also to Dr. R. V. Andrews for making the plasma cortisol estimations in the wolf cubs which are included in this report.

Much of the material incorporated in this text, some of which has been published separately as short articles in scientific journals, constitutes an integrated series of studies which provides a valuable multidimensional view of domestication and behavior.

My research was supported in part by NSF grant GB-34172, PHS grant P 10-ES-00139 awarded through the Center for the Biology of Natural Systems, Washington University, and a grant from the Arctic Institute of North America under contractual arrangement with the Office of Naval Research.

Thanks also to Ms. Wanda Meek for her invaluable secretarial assistance in preparing this manuscript.

Introduction

Over the past 13 years a series of studies on the development of brain, behavior, and socialization in various breeds of domesticated dog have been undertaken (Fox, 1965 and 1971). This research was subsequently refocused on wild (undomesticated) canids, including the wolf, coyote, and red fox, in order to help identify changes in behavior and socialization due to domestication in dogs (Fox, 1971b). Differences among various wild species, in terms of communication and social organization, correlated with behavioral adaptation to a particular niche or set of ecological variables (Fox, 1975).

These observations led to the question of the adaptability of domestic dogs to "revert" to the wild, to become feral and live independent of man. The reversible and irreversible effects of domestication might then be demonstrated. A study of feral urban dogs was undertaken to answer some of these questions. This study underlined the tragic consequences (to the dog) of irresponsible ownership, and several social, ecological, and public health hazards were demonstrated and publicized.

Earlier studies of behavior development and socialization of the dog provided the basis for a number of publications to educate the general public and thereby improve the relationship between man and dog through understanding (Fox, 1972). Following some, but considerably less, research on the behavior development of cats and an extensive survey of the literature, similar material of an educational nature was prepared and published for the same purpose (Fox, 1974).

Research on the effects of early experiences (e.g., social isolation, restricted socialization) on later behavior served to establish some parameters for interpreting and treating abnormal behavior and emotional reactions in adult dogs and other domesticated animals (Fox, 1968 and 1972). This, coupled with nationwide consultations generated via the media (radio and television broadcasts and syndicated newspaper and magazine articles) and referrals of cases by veterinarians in private practice, produced data on a wide range of behavioral anomalies in pets, and often related problems in the pet–owner relationship.

The phylogenetic origins of the dog, established long before man's intervention, are still evident today, together with the effects of 10,000–14,000 years of selective breeding during its period of domestication. As the various needs and life-styles of people change, so the relationship between pet and owner and pet and home environment changes. Such contemporary influences represent the more dynamic and labile aspects of the process of domestication, an understanding of which is important for the well-being of animal and man alike.

Some of the more general aspects of domestication first are reviewed from an ethological viewpoint and various concepts and principles outlined. Given the complexities of the domestication process and the lack of historical records, an objective study of the effects of domestication in any given species must be conducted at four levels: *comparative* behavioral and socio-ecological studies of wild and domesticated canids; genetic studies of hybrids of wild and domestic species; socialization and effects of early experiences; the development of social relationships in hand-raised wild and domesticated species. The strong emphasis in these studies on the behavior of wild canids, which may seem irrelevant or superfluous, actually provides the only template for identifying and analyzing the possible effects of domestication in the dog. Additional aspects of behavior development relevant to the complex phenomenon of domestication are also reviewed, and some unexplored potentials for future improvement of livestock and pets alike is discussed.

I

Effects of Domestication in Animals: *A Review*

Introduction

Domestication, according to a definition outlined by Hale (1962), is a condition wherein the breeding, care, and feeding of animals is more or less controlled by man. The process of domestication involves some biological change in the animal over a period of generations. Such change is apparent in the characteristic behavior patterns of the species and in the morphology and physiology of the domesticated strains.

Some terms commonly associated with domestication and domestic animals must be clarified. The *taming* of animals is frequently misconceived as domestication. By definition, taming is the process whereby an animal's tendency to flee from man is gradually eliminated. This process, while not true domestication, may actually be one of the first steps. If an animal, after domestication, reverts back to a wild state and natural habitat, it is called *feral*. This is quite possible for most domestic species since the process of domestication has not removed natural survival instincts but rather has altered or suppressed certain of these instincts over generations in order to adapt the animal to its domestic environment.

3

Table I.

Under natural conditions

1. Emergence of a new character or character complex making it possible to occupy previously unoccupied ecological niches.

2. Explosive adaptive radiation into all available ecological niches (carnivores, herbivores, etc.).

3. Darwinian selection between species.

4. Adaptive radiation within species.

5. Increasing specialization within the ecological niche.

6. Specialization by minor or nonadaptive (neutral) specialization.

7. Cessation of discernible evolutionary change—highly specialized in an unchanging environment (*e.g.*, horseshoe crab); extinction if unable to adapt to the changing environment.

Under domestic conditions

1. A shift in human culture provides a new ecological niche for those species with favorably adapted characteristics.

2. Explosive domestication of many forms to serve human culture (scavengers, milk producers, draught animals, etc.).

3. Darwinian selection between species.

4. Divergence within domesticated species toward different specialized functions (*e.g.*, among dogs, cattle, etc.).

5. Increasing improvement along a straight line for a behavioral function (increased milk production, speed, etc.).

6. Development of breeds or varieties by nonfunctional (morphological) diversification and imposed reproductive isolation.

7. Some forms apparently do not respond to further selection; the condition is likely to be transient, or the stock replaced by a line which will continue to respond.

The factor of time over generations is a vital aspect of the process of domestication. Domestication is an evolutionary process resulting from changes in the selection pressures on a species or population created by an artificial environment, with release from the competition for survival characteristic of a natural habitat (Allee

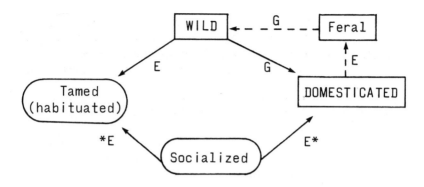

E = Experiences, pre & postnatal

G = Genetic influences (adrenal-pit; neoteny; pseudospeciation

* Infantilism; dependency

Figure 1. *Schema of interrelated factors in the taming and later domestication of a wild animal. Socialization and related environmental/experiential factors and genetic selection are entailed in producing a domesticated phenotype.*

et al., 1949). The significance of such environmental influences are elaborated upon in this book. Such conditions produce changes in the gene structure of a population and differential reproduction of genotypes over generations which could result in eventual speciation. Hale (1962) compares the general evolutionary sequences in animals under natural and domestic conditions. (See Table I and Figure 1.)

History of Domestication

Though there is little agreement over the history of domestication, scholarly inquiry has produced a number of interesting theories. Zeuner (1963), in observing the different types of social relationships between animal and man, has proposed that domestication occurred long before the economic potential of animals in agriculture was discovered. He outlines five stages of the intensity of domestication, which describe both present relationships between man and animal and the historical development of domestication.

1. Loose contacts with free breeding.
2. Confinement to human environment with breeding in captivity.
3. Selective breeding organized by man to obtain certain characteristics and occasional crossing of wild forms.
4. Economic considerations of man leading to the planned development of breeds with certain desirable properties.
5. Wild ancestors persecuted or exterminated.

With this outline of social relationship, Zeuner (1963) developed a probable order in which species were taken into domestication.

1. Mammals domesticated in the pre-agricultural (Mesolithic) phase.
   a. Scavengers: dog, pig, duck.
   b. Social parasitism: reindeer, sheep, goat.
2. Mammals domesticated in the early agricultural (Neolithic) phase.
   a. Crop robbers: cattle, buffalo, gaur, batteng, yak, pig.
   b. Those systematically domesticated: fowl, hyena, ostrich.
   c. Pest destroyers: cat, ferret, mongoose.
3. Mammals subsequently domesticated primarily for transport and labor.
   a. Domesticated by agriculturalists in the forest zone: elephant.
   b. Domesticated by secondary nomads: camel.
   c. Domesticated by river valley civilizations: ass, onager.
4. Various other mammals.
   a. Small rodents: rabbit (Medieval), dormouse (Roman).
   b. Experimental domestication: hyena (Egyptian), fox (Neolithic), gazelle (Egyptian), ibex (Egyptian).
   c. New World species: llama (American Indian).
   d. Pets: Mouse (modern European).
5. Birds, fishes, and insects (not classified).

Artificial Speciation and Adaptive Radiation

Under any condition, natural or domestic, the major factor producing evolutionary change is isolation of one population of a species from another. In the domestic situation, a population is intentionally isolated by man (whereas isolation under natural conditions is a result of chance). Such isolation occurs on two levels. Geographic or reproductive isolation occurs where a topographical barrier prevents two populations of a species from interbreeding. This leads to divergent adaptation of each population to its environment through new phenotypic expression of the common genotype. If the populations are allowed to interbreed (gene flow) after a relatively few number of generations, speciation will not occur. Should such gene flow be inhibited by geographic isolation for a long period of time (in generations), divergence might occur to the extent of making the two populations reproductively incompatible, and hence separate species. Isolation on this level is twofold. Behavioral isolation is the result of divergence of courtship-mating patterns between populations. Incompatibility may also result from divergence in the characteristic anatomies of the two populations making reproduction mechanically impossible. Most domestic strains are reproductively compatible with the wild counterparts of their species. There are very few species of animals that are solely domestic.

It is obvious, therefore, that domestication is, as is any evolutionary process, the result of an intricate combination of environmental influences coupled with the hereditary potentials of a species. Through manipulation of the specific environment, man can accelerate the evolution of certain characteristics of a species to suit his specific needs. Other environmental manipulations, which may greatly modify the animal's behavior, emotional reactivity, and sociability, will be discussed in Chapter 9.

The Effects of Domestication on Behavior

MAJOR INFLUENCES

With an understanding of what domestication involves, we can begin to look at its various effects. There are a few major factors inherent in the process of domestication itself that directly influence the behavior of a species (Fox, 1968c). The first of these is the change in environments from a natural habitat to the habitat created by man. The basic evolutionary aspects of this change already have been discussed. An animal will respond to the type of domestic environment in which it is placed. An artificial environment, depending on the specific circumstances, can have a great influence on the general behavior, disease resistance, and total productivity of the species. The average zoo, where conditions of captivity have many of these effects, is a good example of this. Different individuals and different species will adapt better to this type of situation than others. A simulated natural environment frequently is easier to adapt to and provides for a minimum of incidence of behavioral abnormalities and stress diseases. Man will create either of these domestic environments to suit his needs (Fox and Walls, 1973).

The second major factor influencing behavior (Fox, 1968b) is the genetic selection of specific strains from a few species of animals originally domesticated. Man selects such strains for a number of general characteristics.

1. *Docility*—this is a heritable trait.
2. *Adaptability* and fitness for different domestic environments.
3. Desirable *characteristics* of *economic importance* which can be enhanced through breeding such as high fertility, rapid growth, efficient food conversion, etc.
4. Fixed *paedormorphic features* which reduce the time span from birth to maturity, but allow a high degree of adapt-

ability because of perpetuation of certain infantile characteristics.

5. Reduction of wild characteristics (especially aggressive and sexually related display structures) such as horns, hair, etc., by *neoteny* (similar to 4 above).
6. *Hybrid* vigor through crossbreeding.

This type of direct (genetic) selection will precipitate a far more specialized deviation from the natural behavioral norm than the changes in behavior patterns resulting from simply a change in the environment. Also, any exaggeration of certain physical characteristics through breeding will be accompanied by adaptive changes in characteristic behavior.

CHARACTERISTICS FAVORING DOMESTICATION

A number of natural behavioral characteristics in animals are very adaptive in facilitating transition from a wild to a domestic state. Hale's (1962) observations are summarized in Table II. A domesticated species might not show all of the characteristics listed on the left side of this table, but it is unlikely that a species showing a majority of the behavioral characteristics in the right column could adapt to a domestic situation.

Group Structure

In a domestic situation, flock- or herd-type social groups with often well-defined dominance hierarchies are much more manageable and easily maintained than a territorial, family social structure. A loose social hierarchy or leader–follower social organization reduces the number of conflict situations that would arise due to the social stress of confinement, and the large social group would have a similar advantage over a territorial species in situations where large numbers must be housed or penned together to provide large quantities of produce. (An exception to this is the domestication of territorial species as family pets.) A reduction in proximity intolerance (social distance) and intraspecific aggression may also have had high selection priorities in domestication.

Table II.

Favorable characteristics

1. Group structure:
   a. Large social groups (flock, herd, pack), true leadership.
   b. Hierarchical group structure.
   c. Males affiliated with female group.

2. Sexual behavior:
   a. Promiscuous matings.
   b. Males dominant over females.
   c. Sexual signals provided by movement or posture.

3. Parent–young interactions:
   a. Critical period in development of species-bond (imprinting, etc.).
   b. Female accepts other young soon after parturition or hatching.
   c. Precocial young.

4. Responses to man:
   a. Short flight distance with man.
   b. Little disturbed by man or sudden changes in environment.

5. Other behavioral characteristics:
   a. Omnivorous
   b. Adapt to a wide range of environmental conditions.
   c. Limited agility

Unfavorable characteristics

   a. Family groupings.
   b. Territorial structure.
   c. Males in separate groups.

   a. Pair-bond matings.
   b. Male must establish dominance over or appease female.
   c. Sexual signals provided by color markings or morphological structures.

   a. Species-bond established on basis of species characteristics.
   b. Young accepted on basis of species characteristics (*e.g.*, color patterns).
   c. Altricial young.

   a. Extreme wariness and long flight distance.
   b. Easily disturbed by man or sudden changes in environment.

   a. Specialized dietary habits.
   b. Require a specific habitat.
   c. Extreme agility.

Sexual Behavior

The most important adaptive trait of a species is its ability to reproduce under domestic conditions. Promiscuous sexual behavior has obvious advantages over the establishment of pair-bonds in domestic situations where certain traits are being bred for, and sires are being used. Established male dominance over females reduces conflict between courtship behavior and aggression, thus normally facilitating mating behavior. Posture and movement as sexual response eliciting stimuli are more stable and adaptive and presumably more important than sexual display structures. The latter secondary sexual characteristics are less necessary for reproduction in captivity since their reduction (through neoteny) does not lower reproductive success in captivity (although it would in the wild). Disruption of the former characteristics may directly affect the breeding capabilities of the species. Other social and developmental influences on reproductive behavior are discussed in Chapter 9. (See also Enders, 1945.)

Parent–Young Interactions

The establishment of a species-bond by imprinting of young has distinct advantages in adapting to a domestic environment. During the critical period of imprinting, young can be separated from parents in groups and raised by humans, thus altering their sexual preferences little and socializing them to man. Precocial young have an earlier opportunity than altricial young for imprinting during the critical period. Acceptance of alien young by some species (allowing transfer of wild young and their adoption) facilitates successful establishment of a species under domestic conditions. Recognition of young by species color markings (or other specific cues) is less adaptive under these conditions because females might kill young of other mothers because of variations in color patterns or overt behavior per se.

Responses to Man

A short flight distance and minimum disruption in the presence of humans facilitates handling and rearing. Socialization with man early in life eliminates or greatly reduces flight and critical distance responses (see Figure 15, Ch. 10.) as does imprinting to man soon after birth in precocial vertebrates. Further details of so-

Table III.

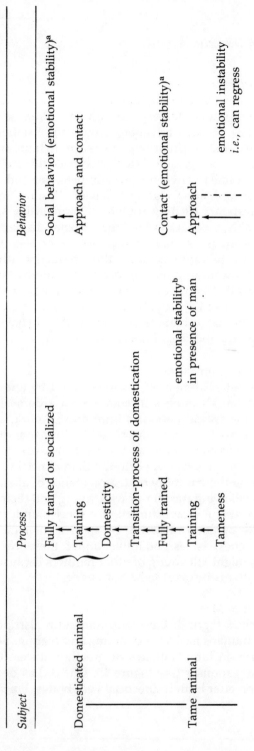

Subject	Process	Behavior
Domesticated animal	Fully trained or socialized ← Training ← Domesticity ← Transition-process of domestication ← Fully trained ← Training ← Tameness	Social behavior (emotional stability)[a] Approach and contact
Tame animal	emotional stability[b] in presence of man	Contact (emotional stability)[a] Approach emotional instability *i.e.*, can regress

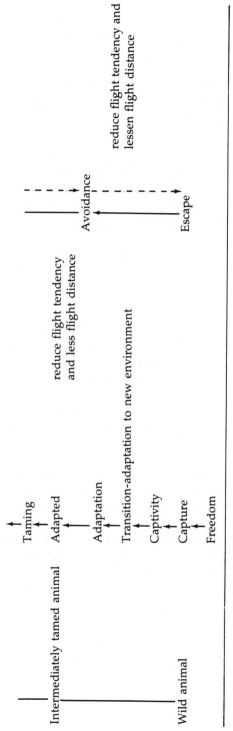

reduce flight tendency and
lessen flight distance

Avoidance

Escape

Intermediately tamed animal

Taming

Adapted

Adaptation

reduce flight tendency
and less flight distance

Transition-adaptation to new environment

Captivity

Capture

Freedom

Wild animal

[a]Dependent upon predictable behavior of human being.
[b]Trainability impaired by delayed socialization.

cialization in wild versus domestic canids is discussed in detail in Chapter 8. Genetic selection for high proximity tolerance and high response threshold and flight distance to novel stimuli may have been a consequence of selection for docility (see also Beylaev and Trut, 1975).

Socialization plays a key role in both the initial phase of domesticating a species and in maintaining domestic conditions with each generation. An animal must be socialized to humans without altering the sexual preferences of the species or causing rejection of the animal by its own species. Constant handling at a young age will socialize wild rats to humans, but the rats must be handled throughout their lives to maintain this tractability (Richter, 1954). Similar "regression" in the absence of continued social contact has been noted in some foxes and coyotes, although individual variability is considerable. Wild adult animals may become habituated (or tamed) to man, but only the more sociable and gregarious (e.g., wolves [Woolpy, 1968]) can become socialized (i.e., emotionally attached) when adult. When young, socialization is usually very rapid but may not endure if that species is by nature nongregarious or relatively asocial (see Chapter 8 for further observations).

Table III, modified from Fox (1968b), evaluates the various stages involved in the socialization of animals to humans.

Food and Habitat Characteristics

"Ability to meet needs for food in man's environment or live on the byproducts of his agriculture is essential for complete domestication" (Hale, 1962). A flexible diet is adaptive under domestic conditions.

Specialized habitat requirements are a disadvantage under domestic conditions as confinement prevents searching for ideal habitats. Thus limited agility is also adaptive.

BEHAVIORAL CHANGES INDUCED UNDER DOMESTICATION

Eibl-Eibesfeldt (1970) views domestication as resulting in generally simpler behavior. Alterations in characteristic behavior patterns under domestication are the result of the various adaptations a species must make to a given set of conditions. Such adaptations, in many cases are learned, and modification occurs in the appetitive sequences of behavior. An example of this might be a change in the manner a domestic animal searches for food. Pigs might learn that food is kept in a feeder and it can be obtained by lifting the cover of the feeder rather than by searching for food in a field or pen. New appetitive behavior leads to the species' characteristic consummatory response, in this case, eating behavior characteristic of pigs, as these patterns of behavior are highly stable.

Adjustments in behavior can also result from simply learning to respond in characteristic ways to new stimuli associated with the domestic environment. Stimulation for the release of milk in dairy cows is frequently switched from manipulation of the udder by calves to the rattle of cages and equipment at milking time (Ely and Peterson, 1941).

The effects of selective breeding are readily observed in the more malleable behavioral traits. As has been discussed, most basic motor patterns resist modification under domestic conditions. Certain of these patterns will combine through hybridization sometimes creating difficulties because of ambivalence in the reactions of the hybrid.

Under domestication, territorial patterns tend to break down. Those species with a greater flexibility in their social structuring generally adapt better to the domestic environment. Hale (1962) points out, though, that some species have a "propensity for divergent social structures." Some species whose social grouping is based on hierarchical structure in a wild habitat become territorial under domestic conditions. Changes also occur in the mating habits of some species where reversion from pair-bond arrangements to promiscuous mating takes place.

Selective breeding shows a marked effect on the frequency and intensity of certain patterns of behavior through changes in response thresholds necessary to elicit certain responses. Apparent elimination of certain of these patterns in domesticated populations is the result of an increase of these thresholds. This has been seen in experiments with a number of species of animals. Extensive experimentation has been done with the Norway rat. It has been shown (Barnett and Stoddart, 1969) that domesticated strains of this species have higher thresholds for aggression in a conflict situation and are less suspicious of new objects, food, etc., as a result of a higher threshold for avoidance behavior and neophobia (Barnett, 1958). Keeler (1970) has shown that selective breeding for coat color in color-phase foxes produces higher thresholds for behavior associated with the fear response.*

Because the domestic environment removes many of the selection pressures of the natural habitat, behavior patterns adaptive to competition for survival are not as adaptive under domestication. Domestication does not remove responses of fright, aggression, etc., from the repertoire of behavior of a species, and occasionally such behavior patterns occur spontaneously as vacuum activities (Fox, 1968b). More usually, however, without reinforcement and no longer being of adaptive value, they may be easily eliminated through selective breeding.

"Purely intra-specific selective breeding can lead to the development of forms and behaviour patterns which are not only nonadaptive, but can even have adverse effects on species preservation" (Lorenz, 1968). Selection for traits of economic advantage to the breeder can result in disruption of the coordination of certain patterns of behavior unless stability of these behavioral traits is bred for simultaneously. This disruption has been seen in certain strains of turkeys bred for extreme breast size. Some males exhibit sexual-copulatory behavior toward piles of dirt, rarely mounting females. Perpetuation of these strains is accomplished by artificial insemination (Hale, 1962). Complete evaluation of many of these behavioral disruptions is complicated by the lack of adequate information on the occurrence of such patterns in wild strains.

Threshold changes induced by domestication have already been mentioned. Occasionally selection for certain behavioral traits leads to undesirable changes in other patterns of behavior. Certain strains of turkeys bred for hypersexuality have shown decreased thresholds for imprinting to man, resulting in the display of sexual behavior directed toward the caretaker!

Another maladaptive consequence of intense selective breed-

---

*In view of the marked differences within dog breeds of individuals having different coat colors, a detailed study of the relationship between coat color and temperament is needed.

ing is the appearance of behavioral phenodeviants in hybrids which are not apparently present in either parent strain. A certain hybrid strain of turkey has been observed denuding many members of the flock spontaneously. Deviant behavior has been seen in mammals where in one case a hybrid strain of female mice is known to eat specific digits from her offspring while cleaning them. This phenomenon, according to Hale (1962), remains basically unexplained.

NEW ADAPTIVE PEAK ACHIEVED UNDER DOMESTICATION

As man domesticates a species he is endeavouring to shift that species from the adaptive peak representing the natural habitat and ecological niche to a new adaptive peak representing the artificial, domestic habitat. The success of this shift is primarily dependent on man's ability to maintain the reproductive fitness of the species. Many species have a number of characteristics favorable to domestication, and the valley between peaks is a shallower one than the transition for species with fewer favorable characteristics. Occasionally, a species well adapted to the domestic environment may lose some reproductive fitness during selective breeding for specific specialized traits. This problem can be dealt with by relaxing selection for the desired trait causing some regression of the trait, but allowing the reproductive fitness of the strain to increase, a new gene pool to establish itself, and thus copulation settling at a new adaptive peak. Hybrids produced from crosses between domestic and wild strains of the same species frequently show combinations of characteristics unfavorable for successful survival in either habitat. Crossbreeding between two or more inbred lines of domesticated animals (dogs or cattle) can result in considerable improvement in performance—a valuable asset afforded by hybrid vigor.

The Effects of Domestication on Physiology

Changes in the environment alter the physiological needs of a species, and corresponding adjustments of behavior occur. Thus, it is important to consider physiological change as a basis for many of the changes in behavior due to domestication.

The hypothalamus has been implicated in changes in the degree of characteristic aggressiveness in wild and laboratory bred rats. As Barnett and Stoddart (1969) have pointed out, laboratory bred rats are much less prone to attack or threaten a strange rat in a conflict situation than a wild rat which will frequently kill a stranger. This "killer instinct" has been "tamed" in wild strains by the injection of methyl atropine into the hypothalamus and surfaced in domesticated strains by a similar injection of various cholinomimetics. Rats are indicative of the general trend toward docility in domesticated animals, and the possibility of hypothalamic change as a cause of this is currently being researched.

Richter's (1954) work with the Norway rat has shown that domesticated strains have a lower characteristic metabolic rate resulting in a lower food and water intake per kilogram of body weight. Coupled with this is a lower resistance to poisoning than is seen in wild rats.

Nervous system changes accompany domestication. Changes in the frontal poles are another factor in the docility seen in domestic rats. Removal of the frontal poles causes domesticated rats to lose much of the savageness of wild rats (Richter, 1954). The sensitivity of various neural structures to hormones may change. These specific changes cause changes in both the physiological and behavioral thresholds of the strain (Hale, 1962).

Beylaev and Trut (1975), in selecting for docility in successive generations of silver foxes, found changes in morphology, estrous cycles, and adrenal responses to stress and ACTH administration. They conclude that "selection for docile, tractable behaviour leads to the dramatic emergence of new forms (phenotypes) and to the destablilization of ontogenesis manifested by the breakdown of correlated systems (adrenal–pituitary, gonadal–pituitary) created originally under stabilizing selection." They present an important concept of *destabilization* where artificial selection can increase phenotypic variance and alter the wild phenotype (structurally, physiologically, and behaviorally) as exemplified by the domestication of animals.

Several of the general and theoretical issues pertaining to the effects of domestication are examined experimentally, with first a detailed analysis of the social behavior of wild versus domestic canids.

II
Socio-Ecology of Wild Canids: Environment and Behavioral Adaptation

Introduction

The canid family contains a variety of diverse and fascinating species which have a global distribution, ranging from the Arctic to the African desert and Indian and South American jungles (Staines, 1975). Some species, such as the red fox, are solitary except during the breeding and infant-rearing season, while others, like the coyote, form more permanent pair-bonds and remain together during the nonbreeding season. A few species, such as the wolf, are even more gregarious and usually stay together as a pack. In these three canid types, it is possible to trace the evolutionary changes that have occurred in each type now specially adapted to a particular ecological niche. The life style (behavior, communication systems, social organization, hunting patterns, and major prey utilized) reflects the social and ecological adaptations that have taken place during the evolution of the many members of this genus.

21

The Distribution of Canids

The domesticated dog has many cousins which belong to the family Canidae (see Figure 1). These include the coyote and wolf, various species of fox and jackal, the Indian dhole, and the Cape hunting dog (Staines, 1975). The doglike "foxes" of South America, such as the maned "wolf," bush dog, small-eared dog, crab-eating fox, and Culpeo are unlike any of the other canids, their uniqueness probably being related to the great length of time that the continent of South America was effectively cut off from the rest of the world. In contrast, the jackals of Africa and Asia more closely resemble the coyote of North America and, similarly, the pack-hunting dhole of India resembles somewhat the Cape hunting dog. These similarities may be interpreted as evolutionary parallelisms and convergences. For example, the coyote and golden jackal may not, in fact, be distantly related, but, instead, have come to resemble each other because of their specialization to a particular niche or life-style.

In the Northern hemisphere, the wolf is widely distributed and may be regarded as the most successful of the pack hunters (the other two pack hunters being the Cape hunting dog and the dhole), being found as far north as the Arctic, as far south as Mexico, and as far east as Asia. There were many races or subspecies of wolf, most of which are now extinct because of man's invasion and exploitation of their hunting ranges and systematic destruction of predator species. Consequently many races of wolves, superbly adapted to particular regions after generations of evolution, are now extinct. Similarly, the red wolf (*Canis niger/ rufus*), once abundant in Texas and Florida, is now virtually extinct. A few specimens still remain in the Texas gulf area, many of which have been crossbred with more tenacious coyotes, so that few, if any, red wolves are left.

The other two species of pack hunting canids are also facing possible extinction. Until it was realized that predation by Cape hunting dogs was beneficial to quality and population control of hoofed animals in Africa, the dogs were shot on sight. (In some areas they are still shot for "disturbing" the herds of big game!) In

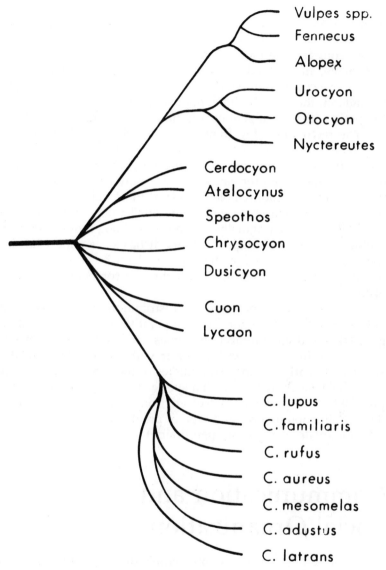

Figure 1. *Schema of taxonomic relationships of canid species based upon behavioral evidence (from Fox, 1975). Note the common origin of wolf (C. lupus) and dog (C. familiaris and other wolflike canids), probably differentiating as separate species prior to man's domestication of the dog.*

India, the dhole is in a similarly precarious position, as are many of the carnivores (see Fox, 1978).

In contrast, the more solitary predators, such as the red fox and coyote, are more adaptable to human intrusion and predation. The elusive coyote numbers at least 40,000 in Kansas alone, while in England, there is a high fox population in Tilbury docks, which competes with the indigenous warehouse cats for rats and garbage.

The raccoonlike dog from Japan, which in the fall becomes very fat and may hibernate in the winter, is another example of a successful canid. Imported to Finland for commercial fur production, it escaped from the ranches and is now wild in many European countries including Poland and Czechoslovakia. The dingo of Australia, which closely resembles the smaller dogs of New Guinea and Malaysia, was originally a domesticated dog brought into the continent by early aboriginal settlers. It has since gone wild, becoming well established as the main predator because there were no other placental carnivores with which to compete (Macintosh, 1975).

Having briefly considered the distribution of canids, we will now look at their behavior. The comparative analysis of social organization and communication signals can give us some insight into the evolution of social behavior within a family of related species. Similarly, comparative studies of the same species in different habitats show how such a species has adapted socially and physically to a particular set of environmental conditions (e.g., distribution, abundance, and type of prey) and differences in terrain (mountain, tundra, or open prairie).

Communication and Social Organization

The canid family comprises an assembly of species that span the entire spectrum from the relatively solitary red fox to the gregarious wolf. Consequently, this family provides a valuable research opportunity for studying the evolution of communication mechanisms in related species differing in social behavior and in the socio-ecological patterns or life-styles that they have evolved. The relatively solitary canids such as the red fox have a simple repertoire of visual signals—tail and body positions and facial expressions. Similar basic signals which serve to increase, decrease, or maintain a certain social distance or proximity are recognizable in the more social canids, but in these species, the signals are more variable and are subtly graded in intensity. Greater complexity of signals and, therefore, greater message carrying potential is also afforded by rapid successive and simultaneous combinations of these signals. Thus, the wolf and domesticated dog can give successive alternate signals of submission and defensive aggression or can combine, simultaneously, signals of submission and greeting. Very similar signals are present in the less social canids, but they often appear at a "typical intensity" and lack these more subtle intensity shifts. The threat gape of a red fox is very much an all or nothing signal, while the wolf has a greater repertoire of threat signaling (Fox, 1971b).

These generalizations may point to the fact that with increasing sociability and sustained proximity, a strong selection priority was placed upon the development of more sophisticated visual signals.

Marler and Hamilton (1966) have pointed out that the more solitary primates have highly stereotyped calls, while the more sociable species have more subtle and variable vocal patterns. A similar generalization may be applied to the canids, the more solitary of which have elaborate and stereotyped calls which are used especially during the breeding season to locate a mate and to drive off rivals. The more solitary the species (which also tend to be nocturnal), the greater is the reliance and priority upon auditory communication because visual signals are ineffectual when individuals are separated by some distance. Wolves and coyotes also have distance-communication signals; their howls, interspersed with barks and yips, serving to bring companions together and possibly to inform rival individuals or packs of their presence. The more solitary canids apparently have not evolved a howl in their vocal repertoire, so that the howl seems to be correlated with more complex social interaction (see also Chapter 4). Wolves and Cape hunting dogs often engage in a mutual greeting and singing (or howling) ceremony before they go out to hunt; such group behavior may serve as a pep rally, bringing all individuals to the same degree of excitement.

The more solitary canids also seem to rely a great deal upon olfactory communication. Odors (pheromones), when deposited on the ground in urine or feces or against particular objects or "scent posts," tend to persist for a long time. The message is, therefore, a much more permanent kind of signal and is effective long after the animal has moved on. Social canids mark out their territories and, like howling, it may serve to inform others of their presence and of their movements in the hunting range which others might use. It would seem logical that the less social canids rely more upon olfactory communication than the more gregarious species, and this is partly supported by the simple observation that the more solitary species smell much stronger than the more social ones. They are more likely to urinate or defecate on a novel odor (of carrion or deer musk), while the more social canids will often roll on such odors, their reward possibly being a lot of social investigation when they join up with their companions later (Fox, personal observations). Olfactory, as well as auditory, communication would be essential in order for the male to locate a receptive female during the breeding season. In the more social canids, where the receptive female is always with her mate or always with the pack, the selection pressures for effective olfactory and auditory communication would be relaxed.

Social-Ecological Interactions

The communication patterns are related to the type of social organization that the various canids have evolved, which in turn reflects the ecological factors to which each species has adapted or has been shaped by. These ecological factors include primarily the species size, annual and seasonal distribution, and abundance of prey. The type of terrain, climate, and seasonal variation in temperature are also significant, as well as the presence of other pred-

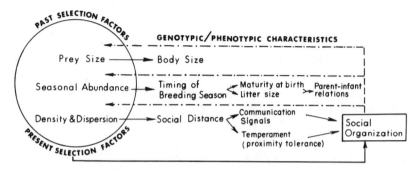

Figure 2. *Schema of interrelated genetic and environmental variables which determine genotypic/phenotypic characteristics in the socio-ecology of canids (from Fox, 1975). Human intervention, as in domestication, can clearly influence these interrelationships at almost every interphase, leading to increasing genotypic destabilization and phenotypic variance in such variables as proximity tolerance, temperament, body size, and reproduction.*

ators of greater or lesser size which occupy competitive or non-competitive niches in the same habitat.

Competitive niches are rare and when competition between predators occurs, the ecological balance may be disturbed. Where there is an abundance of many different prey species ranging greatly in size, a greater diversity of predators can be supported. Thus, wolf, coyote, and red fox, or Cape hunting dog, golden jackal, and bat-eared fox may be found in the same habitat, each in its own particular niche and each occasionally scavenging on the remains of a kill made by a larger predator.* Indirectly, one predator may assist another in the complex food chain by regulating population growth and preventing starvation. For example, smaller predators that specialize on small herbivorous mammals (rabbits and rodents) prevent the latter from becoming too abundant which, if they did, would lead to overgrazing which could

*Interestingly, and tragically, the extermination of the wolf in Scandinavia has resulted in a decline in the wolverine, Arctic fox, owl, and raven, and other animals that relied greatly on the remains of wolf kills as their prime source of food.

cause infertility, starvation, and death of larger herbivores such as deer. The larger predators that depend on the latter would then suffer.

The equilibrium maintained between prey and predator populations is remarkable, the complexities of which are only just beginning to be understood. Wolf packs in Alaska and Isle Royale have remained more or less the same size over the past 30 years, and even in years when prey is abundant, the pack size does not suddenly increase (Haber, personal communication). Many of the factors which regulate pack size remain to be identified, but some have been detailed (Fox, 1973 and Fox et al., 1974). In most canids, the breeding season is timed so that the birth of young will coincide with the birth of prey species, at which time food is most plentiful. The prey tend to give birth all at the same time, such reproductive and parturient synchrony possibly enhancing survival because the predators would eat their fill and be satisfied. If the birth of young were not synchronized but instead extended over a longer period, predators would take a far greater toll on the newborn before the latter were able to fend for themselves.

Social Organization

There are three basic patterns of social organization in the Canidae. In Type I, such as the red fox, only a temporary bond is formed during the breeding season between male and female (Burrows, 1968). The male may stay with the female and assist in providing food for the young, but the litter is usually deserted by the parents around 4–5 months of age. No stable hierarchy or peck order develops in the litter, so that, in the absence of strong social bonds and other group-cohesive forces such as leader–follower relationships, the litter breaks up and individuals go their separate ways as solitary hunters.

In the Type II canid, such as the coyote (Gier, 1975) and golden jackal (Van Lawick and Van Lawick Goodall, 1971), there is a more or less permanent pair-bond between male and female, and their offspring may remain with them until the following breeding season. They may then be driven out of the parents' territory, but occasionally when there is an abundance of food, and therefore reduced competition, some yearlings may stay and even assist the parents in tending for the next batch of offspring. Such occurrences are rare. More usually the litter disperses, and each member seeks its own fortune outside of its natal territory. How early familial relationships influence later social interactions between Type II canids sharing the same hunting range remain to be evaluated.

The most detailed studies of the social dynamics in Type III canids have been done on the wolf (Mech, 1970). There is still some doubt about the origin of the pack, but the general consensus is that the pack consists of related individuals of various ages. There is a binding leader–follower relationship between subordinate wolves and the lead or alpha wolf. A dominance order is seen among the females and the males, the alpha, or most dominant male, serving not only as a leader and decision-maker but also as a "policeman" who often intervenes to settle disputes that flare up in the ranks. There are pack rituals that may serve to maintain and reinforce social bonds; subordinates will affectionately and submissively greet and mob the leader, while the leader may "present" the pack with some token food object. Pack members will assist a mated pair in feeding the young and in guarding them when the pack is away on a hunt. As the young mature, they may be recruited into the adult pack and essentially fill vacant places afforded by the death or departure of others. If such vacancies are not available and food is scarce, mortalities will be high. Occasionally such a large pack may split up and the young, their parents, and one or two other adults may move into a different hunting range or remain in the same locale while another segment of the pack moves out. Such emigrations would only be possible where neighboring packs are not in posession of adjacent territory and hunting range. The factors which regulate pack size over generations of offspring remain to be identified. It is known at least that every adult in the pack does not breed and that there is a kind of social control of breeding.* Usually only the dominant female breeds; she subordinates other females and will intervene if they show any overt sexual behavior to the mate of her choice (which may or may not be the alpha male) or to other males.

---

*Group vocalizations in wolves and coyotes may have some social feedback (i.e., epideictic function) to regulate reproduction.

Figure 3. *Interspecies differences in intraspecific aggression.*
Coyote cubs are extremely aggressive while wolf cubs of the same
age engage in sustained bouts of play. Facing page: *At a later*

age, pairs of coyotes and wolves react differently to an introduced stranger, coyotes threatening and intimidating (top), wolves greeting and investigating (bottom right).

Temperament and Behavior

What determines whether or not a young fox or wolf cub will live a solitary or group-oriented life? It seems that much is determined by the basic temperament of the species, as well as by the presence or absence of the group-cohesive forces alluded to earlier. We might postulate that a socal drive independent of a sexual drive in the Type II and Type III canids keeps individuals together outside of the breeding season. Recent research on young wolf, coyote, and red fox cubs supports the notion that temperament is also a major determining factor (Fox, 1975). Young foxes, from an early age, show great proximity intolerance; each one is an individual and each is invariably very confident, inquisitive, and highly efficient at killing prey. In wolf litters, there is a much greater range in individual temperaments, ranging from the shy, timid type (who are often too afraid to attack small prey) who tend to be the subordinates to the confident and outgoing type. The latter are the dominant wolves who are very inquisitive and kill prey at an early age. They also act as leaders or initiators for their less outgoing littermates. Such a spectrum of individual differences, coupled with group-coordinated behavior, social facilitation of actions, and a strong tendency to follow the leader, insures pack formation.

There may be a strong selection primarily for such heterogeneity of temperaments (behavioral polymorphism) within wolf litters, while in the Type I canid, selection has been for a greater homogeneity of temperament. Litters of coyotes so far studied, as would be anticipated, lie somewhere between the red fox and the wolf in that they show considerable proximity intolerance and aggression toward each other, have little group-coordinated behavior and leader–follower tendencies, but are capable of forming a fairly stable dominance hierarchy.

Development of Aggression and Social Bonds

Wolf cubs begin to play around 3 weeks of age, and their bites, unlike coyotes and foxes, are gentle and controlled. They rarely fight—usually to establish dominance relationships around 5–8 weeks of age, especially in those individuals having similar outgoing temperaments. In contrast, the fox species, and coyotes and jackals, are very aggressive, and fighting often occurs *before* they actually engage in play. At 3 weeks of age the bite is not as inhibited as in the wolf, and they seem to have to learn to control it (Fox, 1971b).

When first put together, domestic dog pups that have been raised apart from each other in social isolation until 12 weeks of age play for only brief periods. Play bouts are broken by one biting its partner too hard. After 3–4 days, though, play bouts are more sustained because they have now learned to control the bite. Therefore, social *experience,* as well as genetic influences, contributes to the control of bite intensity.

Social relationships are based primarily upon who is dominant, and dominance relationships are usually firmly established in coyotes and jackals by 4–5 weeks of age. By this age, they are more likely to play, the subordinates often showing more play-soliciting than their more aggressive littermates.

Foxes, although very playful by this age, do not form close affectional ties with each other, and play often ends in a fight. Fox cubs rarely engage in group activities—each one is a confident individualist. Their egocentric temperaments (high proximity intolerance) and loose social ties in infancy suit them well to their solitary life; by 5 months the litter splits up, each one going its own separate way. The parents usually leave the young, and at this time the bond between the parents is also broken. [Occasionally the bond may break after mating, and the vixen raises the cubs herself (Burrows, 1968).]

In contrast to the fox, a litter of wolf cubs has a greater range of temperaments. One or two are dominant and outgoing leaders, others are followers, and some are very timid and dependent. Eskimo hunters state that some wolves never kill prey—perhaps these belong to the latter category. Research on several litters of wolf cubs at the Naval Arctic Research Laboratory at Point Barrow, Alaska has confirmed this polymorphic heterogeneity of temperaments in wolf litters (in contrast to the monomorphic temperament of more homogeneous fox litters [Fox, 1975]). This phenomenon in wolf litters may facilitate pack formation; all cannot be leaders and there must be followers. A greater range of temperaments would facilitate division of labor, of activities or roles in a social group comprised of individuals of different rank, sex, age, and having different temperaments and allegiances. Dr. David Mech (personal communication) has been studying several packs of wolves marked with radio collars in Minnesota and has also tracked several loner wolves. These are not all old outcasts or low-ranking youngsters; some seem to be young dominant wolves that simply cannot integrate with the pack in which there is a firmly entrenched leader.

Fox and Andrews (1973) have recently found significant physiological differences, as well as behavioral differences, in wolf cubs of high, low, and intermediate rank. High-ranking wolf cubs and yearlings have a greater sympathetic tone (higher resting heart rates) and are both behaviorally and physiologically more active than subordinates. These physiological differences discovered in wolf cubs may be the basis for differences in temperament or personality. It remains to be seen if, when a wolf experiences a change in rank, there is concomitant physiological change. For example, would a low-ranking cub with a low resting heart rate have a higher heart rate if it experienced an elevation in rank? Candland et al. (1970) did find that heart rate changes with a change in social rank in groups of chickens and squirrel monkeys.

Foxes together in captivity show little group-coordinated activity and do not follow a leader like the wolf. This innate tendency to follow may be a major reason why wolf cubs stay in a pack. Another reason may be the maintained responsiveness of wolf cubs to their parents (and vice versa). Fox and coyote cubs show a decreasing responsiveness to their parents as they mature, while wolf cubs continue to be highly responsive. A transference may occur where the young wolf reacts to the leader (which may or may not be a parent) in the same way in which it behaved toward the parent as a cub.

The temperaments of young foxes and wolves seem well fitted to their particular life-styles. If foxes were to hunt in packs, one mouse or rabbit would not go far. Since their major prey is small and widely dispersed, their temperament and solitary method of predation are superbly adapted to their particular ecological niche. Wolves, in contrast, cooperate in hunting and are able to transcend their body size limitation by cooperating to secure prey many times larger than themselves. Wolves that prey mainly on deer tend to be smaller in size and in smaller packs than other races that hunt moose, for example. When food is scarce, the wolf pack will temporarily split up, and in some areas where food is scarce, as in Mexico, wolves are rarely seen in packs. In Italy, they do not hunt in packs but are mainly solitary village scavengers (Zimen, personal communication). Such regional ecological differences have contributed to the evolution of various races of wolves.

The coyotes represent a canid type intermediate between the fox and the wolf. They neither form permanent packs nor live a relatively solitary existence but instead tend to maintain a permanent pair-bond. Litters of coyotes show a greater range of temperament than the red fox but less variants than in the wolf, i.e., oligomorphic. They do not disperse until around 11 months of age (i.e., around the time that the parents come into breeding condition). Again, their temperament, greater proximity intolerance for each other, and relative lack of group-coordinated and leader-focused behavior compared to the wolf accords well with their social ecology. They hunt mainly small prey (rabbits, small ungulates) and usually cooperate in pairs. Occasionally when there is an abundance of food, the young, or some of the litter, may stay with the parents and help them rear another litter. This is a transitional type of social behavior more typical of the wolf and has also been reported by the Van Lawicks (1971) in the golden jackal. Coyotes and jackals under certain ecological conditions may therefore show some of the social patterns of wolves. This indicates that their capacities are more flexible than in the red fox, which, even when food is abundant, still tends toward a solitary life. These observations warrant further experimentation to evaluate the genetic and social (experiential) determinants of behavioral adaptation to the

ecology. It would seem that the fox is genetically limited to one particular life-style and ecological niche, while the coyote and wolf are
more flexible. Kummer (1971), in his field studies of various races
and subspecies of baboons, poses the same question.

In conclusion, the red fox, coyote, and wolf represent three
basic canid types; Type I, solitary, with temporary pair-bond; Type
II, permanent pair-bond, with longer interaction with offspring;
Type III, pack-forming, with strong allegiances and pack affilia-
tions and with dominance hierarchies in both male and female
ranks. Under certain sets of ecological conditions, the Type II canid
may show some of the patterns of Type III, while Type III may
adopt a Type II pattern as exemplified by the Mexican and Italian
wolves.

It is not too premature at this stage to hypothesize that temperament seems to be strongly correlated with the ecology and life-
style of the individual and that past social and ecological factors are
responsible for the selection and continuation of temperament
types in the various wild canid species. Relaxation of such selection
factors in domestication might therefore lead to greater variations
in species-typical temperaments;* some breeds of dog resemble
foxes, coyotes, or alpha wolves (e.g., nongregarious, aggressive
terriers), while others are more like middle- and low-ranking
"omega" wolves (e.g., gregarious beagles, often lacking any social
dominance hierarchy).

Domestication

The origin(s) of the domesticated dog and the various breeds is
unknown (Fox and Bekoff, 1975). There may be wolf and even
coyote and jackal ancestry (or perhaps these bloodlines are only in
some breeds)—these three wild species all will produce fertile hybrids when bred with domesticated dogs (Gray, 1954, and Kolenosky, 1971). These may have been added to the genetic diversity
wrought by 10,000 years of domesticating a dingolike dog ancestor.
Certainly the range, variation of size, temperament, and specialist

---

*Which leads to Lorenz's now recanted view of separate jackal and wolf ancestry for
"one-mannish" and gregarious breeds of dog, respectively.

abilities of the various contemporary breeds is an incredible example of what can be done under intensive artificial selection. To see a naive sheep dog pup playing with a companion and showing all the basic actions of sheepherding, including blocking, turning, and driving is a dramatic case of genetic engineering. Certain behaviors seem to be at a lower threshold and are more easily elicited in some breeds than in others (e.g., herding, retrieving), and they are therefore easier to reinforce in training the dog. Other actions have a greater amplitude or intensity in some breeds; the pointing of bird dogs, for example, is more exaggerated than in terriers, while the action to attack or chase has been truncated in pointers (at the juncture of "orientation" in the temporal sequence). Terriers show a strong tendency to seek out (i.e., appetitive drive) suitable objects to chase or attack, whereas a sheep dog may guard the same object. These are examples of *instinct enhancement* which has led to various breeds of dog fulfilling particular roles for man.

The dog differs greatly from other wild canids, not in overt behavior per se, but in its vocal repertoire (see Chapter 4) and also in its sexual and social behavior. In order to successfully breed many dogs, the trait of monogamy or specific mate preference evident in wild canids has been almost eliminated. A desirable stud dog that is only attracted to one particular female would be of little value, and vice versa. Some dogs still show this ancestral trait (Beach and LeBoeuf, 1967); female beagles and Boston terriers, for example, if given the opportunity, have been known to accept only certain males.

Wild canids only have one breeding season per year, which is timed so that there will be an abundance of prey for the young. Domestic dogs, with the exception of the Basenji (which has a photoperiodically controlled single annual estrus [Scott and Fuller, 1965]), have two and sometimes three heats per year, and this trait is inherited. Artificial selection for this trait would certainly enhance production. Production would also be enhanced if sexual maturity was attained earlier, and this is indeed the case in the domestic dog. It attains sexual maturity as early as 6 months, in contrast to 2 years in the wolf and coyote. Here, a combination of genetic selection and improved nutrition may be involved. Male wild canids produce little or no sperm outside of the breeding season, while domesticated dogs are always potent. Although this again enhances productivity, it can become a serious social problem (see Chapter 11). Dogs, then, selectively bred for utilitarian purposes are sexually promiscuous, precocious, and prepotent, and such activities can be afforded since the natural ecological restraints no longer operate.

Experiences with hand-raised canids have shown that coyotes and wolves, especially, become increasingly fearful of strange objects or to a change in their familiar environment around 4–5 months of age. In the wild, this environmental fear may be a consequence of exposure learning (or imprinting to the home range). It would be highly adaptive for a young animal to recognize any change in its familiar territory which could mean danger. In contrast, the dog does not normally manifest such behavior, presumably because in the protective domestic environment which is constantly changing, such behavior would no longer be adaptive and would make such dogs difficult to handle as well as interfering with performance. An analogous fear of strangers and the capacity to develop social relationships later in life will be discussed in Chapter 8.

Wolf cubs show a greater tendency than coyotes to maintain the infant attachment in that they invariably remain bonded to their human "parent" throughout life. Wolves also seem to have a greater capacity to develop secondary social relationships with strange people later in life than other wild canids, but this social potential is still much less than in the dog. The parent bond in the wolf may be transmuted to a leader bond as the wolf cub matures. A similar transition occurs under normal conditions in the relationship between a dog and its master as the former matures. Less gregarious carnivores, such as foxes, which are relatively solitary in the wild and break away from their parents, also will grow away from their human foster parents; the infant bond is not often transferred into a social bond (leader-follower relationship) with maturity.

Domestic dogs may well have originated from a packtype ancestor, since their socialization patterns are closer to the Type III canid than the Type I or II forms. Studies of feral urban and rural dogs (see Chapter 3) reveal their capacity to form packs under optimal environmental conditions, again supporting the view of a Type III ancestry. Because of suboptimal environmental conditions, pack formation is not possible for the dingo. In this species, presumably, unidirectional selection over generations of feral living has resulted in a temperament and social organization of the Type II category (Corbett and Newsome, 1975). Dogs, like wolves, also respond to a leader in adulthood, and this tendency is easily transferred into the home environment where the dog becomes a part of the household "pack" and is ideally subordinate to the master of the house. But this is not always the case, and some of the problems of socially maladjusted delinquent dogs are discussed in Chapter 11.

In summary, a lot can be gleaned from studies of wild canids in attempting to elucidate what effects domestication has had on the dog. This brings us to an important variable, namely, the process of socialization and critical and sensitive periods in development. Some analogies will be subsequently made between the interpersonal relationships of pet and owner and child and parent (e.g., overdependent, overpermissive) which may give rise to a variety of very similar emotional disorders in both dog and child.

Behavior patterns and communication signals in these canids are inherited (in that they develop independent of social experience early in life). Evidence has been presented here to support the notion that the inheritance of certain temperament characteristics or traits (which are most adaptive to a particular life-style or ecological niche) determines whether the animal will be solitary or gregarious as an adult. Socialization during early life also contributes significantly to the formation of family bonds. In other words, by applying personality theory and developmental psychology to the ethology and ecology of canids, a more complete picture of their behavior in relation to the environment can be drawn. A knowledge of the natural history, social behavior and development of each species is an essential prerequisite, and it is unfortunate that only a few studies have so far been completed.

In conclusion then, the truism that the organism is its environment is confirmed in terms of genetic preadaptations which determine sociability, temperament, and adaptability to a particular prey–predator ecological niche. Relaxation of such selection pressures under conditions of captivity—of the domestic environment per se—as well as artificial selection for particular physical attributes and psychological traits has occurred in the domesticated dog. To understand what changes have occurred in the behavior and psychology of *Canis familiaris,* the variables of human influence must be considered. These are discussed subsequently in Chapter 11. At this stage, some preliminary comparisons between wild and domesticated canids are made in relation to undesirable wild traits and desirable traits associated with sexuality, reproduction, and socialization (see Table I).

Summary

Differences in behavior, including patterns of communication, social organization, interpersonal relationships, and hunting behavior in various canids are described. Also species differences in the development of aggression and the significance of individual differences in temperament are detailed. These behavioral differences between members of the canid family reveal evolutionary changes related to adaptation to a particular set of environmental conditions. Eco-specialization, or socio-ecological adaptation, to a particular niche is manifest by a particular life-style and typologically distinct array of behavioral characteristics (temperament). These findings are relevant not only to understanding the evolution of ecologically adaptive behavior patterns, but they also provide a basis for interpreting some of the behavioral consequences of domestication in dogs, where both genetic and environmental changes have been effected by man.

III
Behavior and Ecology of an Urban Feral Dog Pack

Introduction

Much research has been done on the behavior genetics, ontogeny, and ethology of the domestic dog *Canis familiaris* (Scott and Fuller, 1965; Fox, 1971a, 1972a). Most ethological studies have been undertaken in captive conditions, and the question arises as to what the normal environment for studying the domestic dog really is. After approximately 10,000 years of domestication, the dog still retains the capacity to become feral. Whereas many dogs are confined in the home, others have free access to the neighborhood, be it rural or urban. These are designated as free-roaming dogs, as distinct from those feral dogs that have no home and support themselves independent of human assistance. In some neighborhoods, though, people will put out scraps and even commercial dog food for what they regard as strays, a random assortment of feral and free-roaming dogs. Beck (1973) and Nesbitt (1975) have studied both ecotypes in urban and rural environments, respectively.

The study discussed in this chapter focuses on the behavioral ecology of a trio of feral dogs that operated as a pack or social unit in an economically depressed area of St. Louis, Missouri, a city of some 600,000 inhabitants. Several derelict buildings in the neighborhood provided shelter for such dogs.

In an initial survey of several areas of the city, feral dogs were

Figure 1a. *Typical free-roaming feral urban dogs: (A) using available environmental resources, (B, C, and D) showing varying degrees of emaciation and disease, (D) being in a terminal state.*

seen, many of which, by virtue of their condition and behavior, were easily distinguished from free-roaming dogs. The latter were usually in good physical condition, tolerated proximity of strangers up to 4.5–6 m (15–20 ft), and when approached further would run off and seek refuge in their open backyards which were usually no more than 91 m (110 yd) away (see Figures 1a and 1b). From the security of their presumed home sites they would bark at the investigators. Feral dogs showed a much greater flight distance, were more elusive and harder to follow. Sometimes they were flushed out of abandoned buildings which showed signs of canid occupancy (feces, shed hair on old mattresses and carpeting, chewed food cans and cartons and toys, such as a chewed ball or stick). Putrefying and mummified carcasses of dogs were also found in some of these buildings. The physical signs characteristic of many feral dogs were emaciation and skin lesions (probably mange); some were in such poor condition that they were unable to run

Figure 1b. *Free-roaming house dogs: (A and C) foraging for garbage, and (B) leaving home territory for early morning foraging and social interaction in neighborhood.*

away when first located. One was seen in the company of three free-roaming dogs and over a period of 6 weeks became weaker, more emaciated, and the skin lesions more extensive until it eventually succumbed. Of course many of the dogs initially seen could not be classified into feral or free-roaming until we concentrated the study on one particular neighborhood. The neighborhood chosen was determined by the fortuitous discovery of a group of three feral dogs that were spotted one morning and followed into an abandoned house in which dog signs (feces, hair, play objects) indicated that they had been there for several months.

Materials and Methods

The study began in late March, 1973, becoming most concentrated from May through July, with sporadic observations through February, 1974. Most observations were done from an automobile since the subjects were wary of pedestrians, and any attempt to follow them on foot evoked the flight response. After 2 weeks, the three feral dogs would allow closer proximity while the observers remained in the car and would walk within 0.6 m (2 ft) of the car when we were parked. They soon became habituated to our following them provided we kept the car at a distance of 6–9 m (20–30 ft).

The pack, or trio, consisted of one tan shorthaired female mongrel (F) approximately 16 kg (35 lb) weight. She had elongated teats, a sign of repeated litters, and she was probably the oldest dog of the trio. The other two dogs were male, one a yellow German shepherd (Alsatian) (Y), weighing approximately 32 kg (70 lb), and the other a large mongrel (X), with medium-long brindle hair, weighing approximately 29 kg (65 lb). The latter had a permanent hip injury, causing intermittent lameness in the right hind leg. All of the dogs were obviously mature (see Figure 2). All were in poor physical condition at the inception of the study, the shepherd being the most healthy of the three. The other two had extensive skin lesions, probably mange infestations, but by July, all three were clear and their general condition much improved (possibly a beneficial effect of sunlight). Fecal analysis revealed heavy infestations of hookworm (*Ancylostoma* and *Uncinaria*) and whipworm (*Trichiuris*). Although tempted, the observers provided no food or medication and in no way interacted with these animals during the course of the study.

Two, and often three, observers followed the animals on 4- to 6-hr shifts. A total of 90 hr of observation were completed, covering all periods of the day and night. The following data were collected:

1. Daily Cycle of Activity.   The time of day when active in the area of study, either foraging in alleys, traveling, resting in the open, or hunting in the park was recorded. Much time during daylight hours was spent in one of two abandoned houses. Later in the summer the cool basement of one was preferred. The other house, an upstairs apartment, had a conveniently situated small dog door cut into the back door; the leather flap indicated when one or more of the dogs was in or out of the house by being tucked in or out.

2. . Population Activity Index.   Head counts were made at various hours of day and night of free-roaming dogs in the vicinity of the trio's home range. This provided an activity index that could be compared with the activity patterns of the feral pack which might, for various reasons (see discussion), be different.

3. Movements and Home Range.   The range and movement patterns of the trio were plotted on a survey map of the area.

4. Marking Behavior.   The frequency of marking and context (presence of strange dog or marking after another member of the trio had marked, i.e., marking over) were recorded. Scraping after marking was also noted.

5. Leadership.   As the trio traveled, the one to take the lead at intersections and road crossings was recorded.

6. Miscellaneous Social Interactions.   Notations were made as to the frequency and type of interactions among the trio and between them and other free-roaming neighborhood residents and feral dogs that occasionally came through.

7. Pack Cohesion (Fission and Fusion).   The openness and stability of the group of interindividual relationships among the three

Figure 2. (a) Y leaving "flaphouse"; (b) F and X following him;
(c) Y emerging from basement of second house; (d) trio crossing
busy highway at 8:15 a.m. after park "hunt". Facing page:
(e and f), female F marks in park, 2:00 a.m., and X marks over; and
(g) trio resting in the open near second house, 12:30 a.m.

dogs were determined by (a) recording how often any single member left the group and for how long, and (b) how often other dogs joined the trio and for how long.

8. Hunting Behavior.   In the park, the dogs engaged primarily in chasing squirrels. The number of observed chases was recorded and their hunting strategy studied.

9. Other Dog Groups.   In addition to the head count, the frequency of occurrence of dogs in groups of two or more was recorded. This would give an index of the gregariousness of free-roaming dogs and would possibly disclose what factors lead to dispersion and aggregation.

Additional observations were made in November–December, 1973 to test the possibility of seasonal differences in activity, since in the early part of this study the dogs did not go to the park. At this time, they would return to one of their homes 1–2 hr earlier than during the summer schedule.

Results

ACTIVITY

In late spring, the trio emerged from their sleeping quarters (usually the second floor flaphouse, i.e., the house with the leather flap covering the cut in the door) between 11:15 p.m. and 12:00 a.m. and remained active until 6:30–7:00 a.m., when they would retire until the following evening. During the summer, they retired later— between 7:30 and 9:00 a.m., usually to the cool basement of the second house, or they would move to this basement from the flaphouse sometime during the day as the ambient temperature rose up to 30°C (86°F). The later returns to shelter in the morning always followed a 1–1½ hr period hunting squirrels in the park, an activity which was not evident earlier in the study. A kill of squirrels was never observed. Activity was greatest then from around midnight to early morning, during which time the ambient temperature averaged some 7° less than during the day. A lower temperature in the evening and early hours of the morning would be anticipated but the buildings and pavements of this humid city give off much absorbed day heat during the night.

The night and early morning activity was interspersed with rest periods of varying duration (see Figure 2); the trio never returned to the cover of its houses during this time (even during heavy rain), preferring to sleep out in the open, on lawns or porches of occupied houses. When active, they would systematically forage for food in the three alleys and in backyards which opened onto these alleys (see Figure 3). Interactions with other dogs occurred during these times and also in the park, and these will be described later.

Interestingly, the head counts of free-roaming dogs which gave a good index of the activity of other dogs in the neighborhood did not follow the pattern of the trio during the day and early evening (see Figure 4). Whereas the trio were under cover during the day, free-roaming dogs showed two peaks of activity— between 9:00 and 10:00 a.m. and around 7:00 p.m. The dramatic decline in head counts between 6:00 and 9:00 a.m. corresponds

Figure 3. *Trio foraging 2:00–4:00 a.m. in backyards of house (a and b); in discarded garbate at side of road (c); in trash cans in alleys between houses (d and e); and at a grocery store loading dock (f).*

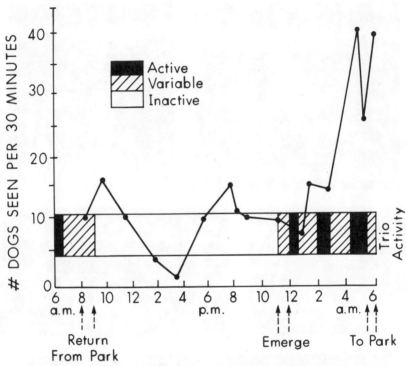

Figure 4. *Activity index of free-roaming dogs in study area (number counted over a ½-hr period). The activity of the feral trio is superimposed for comparison. The latter show no late morning (8:00–9:00 a.m.) or early evening (7:00–9:00 p.m.) activity peaks and are most active during the early hours of the morning. Data averaged from 90 hrs observations.*

with many of the free-roaming dogs returning home, presumably to be fed when their owners arise. The evening peak also correlates with human activity; the streets are full of children playing and adults conversing on their porch steps. The feral trio was never seen out at this time, and it may be concluded that they were avoiding human contact. Most dogs avoided the hottest time of the day between 2:00 and 4:00 p.m.; free-roaming dogs, like the feral trio, were most active during the early hours of the morning, shortly before sunrise (Figure 2).

MOVEMENTS AND HOME RANGE

The trio's main area was located within a low income, primarily abandoned, residential area of about four city streets of 427.5 m (1402.5 ft) long in each direction and 33.5 m (110 ft) representing an available surface area of about 5.7 ha (hectares) (14.2 acres) exclusive of inner buildings. In the mornings, they often used much of an adjacent 55.3 -ha (136.6 acres) park (477.6 m × 1156.7 m) or a total home range of 61.0 ha (150.7 acres). This area is considerably larger than the area used by a group of two feral dogs observed in Baltimore, Maryland (Beck, 1973) possibly indicating a less favorable habitat, i.e., food availability. Home range for many animals is probably influenced by food availability (McNab, 1963). Red foxes, *Vulpes fulva*, were observed to have smaller ranges in areas of greater ecological diversity than those in suboptimal habitat (Ables, 1969). Feral dogs in rural Alabama have home ranges greater than 1,000 ha (Scott and Causey, 1973) possibly indicating less available food. It would be interesting to investigate the hypothesis that all urban populations have smaller home ranges than their nonurban counterparts, which appears to be the case even for people. Human activity and construction may provide for a higher density of food and cover resources and artifacts in the environment that might be used by animals on occasion—all of which would facilitate smaller ranges.

Although the trio kept to a fairly regular time schedule and tended to stay within a predictable home range or foraging area, they twice went into two separate areas where they had never been seen to go before. In one of these, resident dogs were intolerant toward the trio and actually chased them away. This was in sharp contrast to the dogs that lived within the trio's home range; these dogs being more tolerant, some even on friendly terms with the trio (see later). The trio did not seem to be motivated by hunger to go into these new areas since they had recently fed and did not, in fact, forage for food in these new areas. Nor did they seem to be extending their effective home range because they did not subsequently return to these areas. The trio's largest single daily excursion was 1.7 km (1.06 mi) measured point to point after a night of movement across their range.

MARKING BEHAVIOR

The frequencies of marking in various contexts are shown in Table I.
Both males marked more than twice as often as the female (F), who
never, in fact, marked in the presence of a strange dog. Marking in
the alleys of the home range and in the park were at a similar
frequency for X and F, but Y showed a greater frequency of marking in the park than in the alleys. The role of the marker may
change with a change in locale, X being the most frequent marker of
the home site area (alleys), while both males mark almost equally
in the park. Y marked more frequently than X in the presence of a
strange dog and was the most aggressive of the trio toward strangers. The low incidence of scraping after marking in X may be due
to his hip injury; F was never seen to scrape. Marking over, that is
marking where another conspecific has just marked, may be a very
significant social phenomenon in the dog since it was one of the
most consistent findings in our study. Of the 36 times that F urinated, one or occasionally both males were seen to mark over her
mark 30 times (see Table I). Twice, F marked over X. One occasion
served to demonstrate the intentionality behind this behavior:

4:15 a.m. X crosses the road while F pauses to sniff the
           sidewalk. F urinates, then runs to join X but X sees
           her and comes back to mark over. He then crosses
           the road again and continues on his way with F and
           Y following.

X marked over F three times as often as Y; this may indicate a
closer allegiance with the female F. Interestingly, X marked three
times over Y after Y had marked over F. Y was only seen to do this
once (marking over X's mark over F). What stimuli initiated marking in the first place were not determined. Often a horizontal surface was marked (grass clumps, lawn, edge of pavement), but
more often a vertical object was marked (wall, corner of a wall,
tree, fire hydrant, lamp post or trash can). Marking was particularly evident when the trio entered the park early in the morning.

LEADERSHIP

The observed frequency of which individual led the trio when moving within the home range after resting or eating is shown in Table

Table I.
MARKING BEHAVIOR

Animal	Location	Marks alone	Marks with stranger	X over Y/Y over X	Mark over F
X	Park	49 (2 ms)ᵃ	5 (1 ms/1 over Y)	3 (2 after Y marked over F) (1 + scrape over Y marked over F)	7
	Alleys	46 (3 ms)	1	1	8 (1 ms)
	Total	95	6 (5.0%)	4 (3.3%)	15 (12.5%)
Y	Park	45 (10 ms)	(1 mark after X + ms) 10 (2 marks after X) (3 ms)	5 (1 ms) (1 after X and F)	4
	Alleys	29 (2 ms)	5	3	1
	Total	74	15 (14.8%)	8 (7.8%)	5 (4.9%)
F	Park	19	—	—	—
	Alleys	17	—	—	2 (marks over X)
	Total	36 (94.7%)	0	0	2 (5.3%)

ᵃms, marks and scrapes

Table II.

FREQUENCY OF LEADING GROUP

Animal	After resting, eating, and ranging in park	Leads chase in park
Y	1	15
X	12	1
F	37	13

II. The frequency of leading while ranging in the park and while chasing squirrels is also included. Clearly, in the home range, the female (F), was the usual leader. X usually followed F, and Y would then often follow the pair, possibly a magnet effect (see later). Male X tended to lead to the lake in the park where the trio would drink and swim and was also the one to lead the group back from the park (see Figure 5). Male Y tended to lead active chases in the park.

Leadership entailed a number of subtle overt signals, especially eye contact. For example, F or Y would *wait for* X to catch up, often sitting and looking back at him. As soon as they looked back, he would speed up. After resting, F would stand up, walk off a few paces and then turn and look back at X and Y. This *looking back* stimulated them to follow. F would then run or trot a few more paces and look back. If they were not yet following, she would stop, look back and then give an exaggerated head turn in the direction she was heading. The following anecdotes from our field notes will further clarify this subtle communication:

2:50 a.m. Trio sleeping on porch. F gets up, gives a yelp bark and goes down steps onto sidewalk with Y. X sleeps on. F returns 5 min later and gives another yelp bark and a tail wag toward X. X gets up and follows her. F face-rubs and tail-wags with X.

3:55 a.m. Trio sleeping. F stands up, looks at Y. Y immediately sits up. F gets off patio and goes onto sidewalk and looks back at X and Y. She then sets off down the road. Y stands up. F stops and looks back at Y. Y now follows her. X still sleeps. Y looks back at X, then follows F. After 5 min, Y and F re-

turn to patio. F tail-wags at X who is still sleeping. Y
sits close by and tail-wags. X sits up. Y moves off. F
waits by X. Y returns to porch and tail-wags at F. F
goes onto sidewalk and looks back at X. X stands
up, marks twice and Y marks and scrapes. Y crosses
road and looks back at F. F (in middle of the road)
looks back at X, who joins her "at last." She gives a
tail-wag, face-bump and exaggerated head turn to X
and then leads him across the road to join up with
Y. With F in the lead, Y and X follow her to forage.

The following interaction illustrates further the dynamics of
communication between two members of the trio. X is with a resident free-roaming female who is in heat.

8:30 a.m. F moves off and looks back at X. X looks at F, follows her for a few paces, then stops and looks back
at the estrous female. He then returns to this
female. F moves off again a few paces, then stops
and looks back at X. X looks at F and then follows F
back to the flaphouse.

The magnet effect of the trio keeping together is exemplified
by these two observations:

4:30 a.m. F crosses the main highway but X and Y stay on the
sidewalk and do not follow; F returns to them and
all three go off in a different direction.

7:15 a.m. Y crosses the main highway and heads toward the
park; X and F stay on the sidewalk and do not follow. Y looks back and then rejoins the other two.

Further notations on eye contact and orientation will be presented subsequently in the observations on hunting activity in the
park.

MISCELLANEOUS SOCIAL INTERACTIONS

In this category, overt interactions, including eye contact (looking
at), and covert communication of following and waiting for a conspecific, are not included (see section on leadership). Only overt

Figure 5. Trio behavior in the park: (a and b) ranging run, (c) bathing and drinking in lake, (d) treeing squirrel. Facing page: (e) one spatial pattern while looking for squirrels, (f) foraging picnic scraps, and (g) F and Y waiting for and looking at X as he marks tree before leaving park.

social behaviors associated with greeting, play, submission and aggression, and social investigation will be considered. Among members of the trio, a total of only 16 interactions were recorded throughout the entire study. The possible significance of this unexpectedly low frequency of overt behavior is discussed subsequently.

No displays of submission were ever observed between X, F, and Y. On two occasions during feeding, F growled at X and Y, respectively, and both withdrew. Of the 14 remaining interactions, all were friendly and of extremely short duration and included brief tail-wagging, play-bow, face-bumping or rubbing, and nose-pushing toward the mouth of a companion. One brief (5 sec) play-chase was recorded in the park between F and Y. These data are summarized in Table III. X initiated no interactions himself, engaging in one reciprocal face-rub and tail-wag with Y when the latter rejoined the trio after a brief absence. Possibly an indicator of social preferences, the female F initiated more interactions toward X than toward Y, while Y interacted almost equally with X and F.

Interactions with other dogs, some identified as free-roaming residents that shared the same home range as the trio and others as nonresident free-roaming and feral dogs were as follows: Of 33 recorded encounters, six were designated neutral where the stranger approached the trio or passed by and was ignored. Eight interactions were judged as friendly, since reciprocal tail-wagging and, more rarely, play actions were observed (see Table IV). Nineteen other interactions were aggressive or offensive. On one of these occasions, a dog crossing into the park was intimidated by the mere sight of the trio and ran off (none of the trio even looked at the dog), and it was hit by a car. More often the strange dog would be intimidated by Y and would run off. On eight of the aggressive occasions, Y was the main instigator of intimidation toward strangers (see Table IV). On one occasion F supported Y in chasing a stranger.

Table III.
FREQUENCY OF INTERACTIONS WITHIN THE TRIO

	Recipient		
Initiator	X	F	Y
X	—	—	RF[a]
F	A[b] F F F F F[c] (5)	—	A RF
Y	F F F F (4) RF	F F F (3) RF	—

[a]RF, reciprocal friendly interactions.
[b]A, agonistic interactions.
[c]F, friendly interactions.

Table IV.

SUMMARY OF VARIOUS INTERACTIONS WITH STRANGE DOGS (SD)

Threats	*Friendly*
Y barks at SD ♂ and chases; F supports Y, barks and chases. Y then urinates.	Y approaches resident ♂, reciprocal tail-wag and genital sniffing
SD approaches and stands still. Y alerts and looks; SD ? runs off.	Y approaches resident ♂, genital sniffing, then Y marks.
F barks at SD. Y chases SD and then returns and tail-wags to X.	*Y approaches SD ?, investigates, then gives play bow and F prances.
Mange-covered SD ♂ challenged by F. No submission, briefly follows trio. Ignored by X and Y.	Y approaches with tail-wagging and investigates SD ?. F approaches, X follows and SD runs off. Y marks.
*SD ♂ approaches, chased by Y.	*F meets 2 ♂ SDs; social investigation and reciprocal tail-wag but F intimidated until Y joins her and greets the 2 ♂s. X indifferent.
Y approaches SD ♂ who runs off; Y marks.	
SD ♀ approaches submissively; X, supported by F, chases her away.	SD ♀ approaches, gives low tail-wag to Y, who reciprocates with high tail-wag. SD ♀ also tail-wags to F and X. Trio leave.
Y chases approaching SD ?, then marks.	*Trio meet SD ♂, SD ♀ and SD ♂ pup; reciprocal greeting and social investigation. SD ♂ pup joins trio briefly.
SD ? sees trio, retreats though no obvious threat. Y marks.	
*SD ♂ approaches; X tail up, side on threat. Y play-bows to SD, then rushes and bites scruff. SD runs off. Y play-bows to X, then marks.	

[a]Symbols: ?, indicates sex of SD not determined; *, occurred in park. All other observations in alleys.

Both Y and X were seen supporting F in intimidating an approaching stranger.

More rarely, a strange dog would join the trio for varying periods. One resident free-roaming dog, a male, was seen foraging with the trio early in the morning on three occasions. Another morning, a small male pup played with F in the park after meeting the trio earlier in a "morning greeting" (described below) and also briefly joined the trio to hunt squirrels in the park. The most intriguing temporary union of a stranger with the trio occurred one morning. A male dog, about the same size as F, stayed with the trio from 4:42 a.m.–7:40 a.m. while they foraged in the alleys. When the trio returned to one of its houses, he remained outside. The field notes summarize what followed:

> 7:30 a.m. F comes to the front door of the abandoned house (Y and X have retired upstairs). The stranger, a brown male (Br), looks up at F, tail high and wagging, and marks the porch. F comes down the steps and leads Br to the flaphouse. He follows her up the steps. Meanwhile X comes out of the back door of the first house and trails the pair. X and F go into the flaphouse while Br waits on the steps.

His approach/withdrawal behavior in front of the door indicates his ambivalence about entering the unfamiliar house. After waiting 2 min, he leaves. This dog was never seen again by the observers.

Toward the end of the study, a young female dog joined the trio for an estimated 6 days; it was subordinate to F but was accepted by her.

The trio was seen once involved in one of three observed morning greeting ceremonies where several dogs come together and engage in reciprocal social investigation and greeting (see later).

The trio was on friendly terms with two separate dogs who were usually tethered in their backyards. During the early hours, the trio would enter these dogs' backyards, were usually greeted, and would eat out of the dogs' food bowls and forage in the garbage cans in the yards. They would often rest or rendezvous after a brief separation in these yards and use one as a regular crossing between streets. Tail-wagging was often noted between other confined resident dogs and the trio, while foraging in the back alleys of houses.

On three occasions the trio howled to passing police car sirens. They avoided people, especially children, Y having the greatest flight distance. In the park, people were often ignored. Relationships with cats were consistent. All three observed encounters resulted in chasing, and on one occasion a cat was almost caught by F. One rat kill was observed in a back alley; this was the only live prey the trio was ever seen to secure.

PACK COHESION (FISSION AND FUSION)

The frequencies that any individual was not in the trio for any period of time greater than 3 min were as follows. F broke the trio twice for 5 min, once with another nonresident male dog and once to forage alone. X was not with the other two dogs on only two occasions, once when he returned 5 min early from the park and once when he refused to get up and continued to sleep for 5 min before F and Y got him to follow them. Y was the most often away, F and X being seen together in the absence of Y a total of 15 times, indicating that Y may have had a lesser tie to the group. He would often sleep alone in one of the houses. Since we do not know the origins and earlier relationships of these dogs, no further conclusions can be drawn. Usually no greeting or overt interaction was evident when Y rejoined the other two dogs.

HUNTING BEHAVIOR

Shortly after the initiation of the study, the trio, instead of retiring for the day around 6:30 a.m. would go to a park adjacent to their home site for 1–2 hr. (Since they later stopped going to this park, we may conclude that this was a summer schedule activity.) In the park, the trio would scavenge profitably for food under park benches and picnic tables, but most of the time in the park was spent chasing squirrels. On entering the park and marking, they would trot across open space and then break into a "ranging run" beneath the trees. If any squirrels were seen on the ground, they would immediately give chase. The trio would systematically work over different areas of the park where there were stands of trees. Once, and often twice each morning, they would drink and bathe in a large lake in the center of the park for 2–3 min. Bouts of chasing and ranging were interspersed with resting periods, when the trio would lie or sit in three patterns of head orientation (see Figure 5). These patterns optimized the chances of spotting a squirrel and as soon as one dog alerted and oriented (and occasionally tail-wagged), so would the other two. If it then gave chase, the other two would follow immediately, correcting their orientation when they too saw the prey. This later orientation often resulted in one dog bumping into a companion or momentarily running in the wrong direction. Also, while in a ranging run if one dog suddenly ran to one side upon seeing a squirrel, the others would follow at once and then look. If, for example, the center dog in Figure 5e suddenly alerted forward, the dogs on each side would alert in the same direction. If the dog on the far right in Figure 5e alerted in a forward direction and to its right, the other two dogs would turn around quickly and orient to their left.

Chases were usually terminated by the dogs jumping up at the tree in which the prey had sought refuge (see Figure 5d). F and, less often, Y would jump up against the tree, and barking and tail-wagging were frequently noted. On two occasions, squirrels were seen to jump out of the trees up which they had been chased, while the dogs were barking and jumping up at the base of the tree.

Table V.
PARK CHASES

Date	Temperature	Time of onset	Number of chases/time in park
June 8	70°F	5:15 a.m.	6 chases/190 min
June 11	76°F	5:35 a.m.	30 chases/170 min
June 13	73°F (Rainy)	5:33 a.m.	10 chases/180 min
June 15	73°F	5:15 a.m.	10 chases/225 min
July 9	77°F	5:55 a.m.	5 chases/120 min

The dogs expended considerable energy on these mornings in the park. The adaptive significance and bioenergetics of this hunting behavior are questionable since out of a total of 61 chases (see Table V) no squirrel was ever caught. This activity may be reinforcing itself, i.e., like play.

OTHER DOG GROUPS

The occurrence of dogs seen alone and in groups of two or more in the area of study and in adjacent regions is summarized in Table VI. Most dogs were seen singly, although pairs of dogs were common. It was not possible to sex all pairs observed, but where it was possible, either two males or a male and female were together. No female pairs were identified. Six trios of dogs were seen, two of

Table VI.

OTHER DOGS OBSERVED IN THE STUDY AREA

	Number of dogs together						
	1	2	3	4	5	6	7
Resting, moving or foraging	161	30	6	1 (briefly foraging in same vicinity)	—	—	—
With ♀ in heat	—	—	4 groups of 3 ♂s	—	1	2	4
♀ with pup(s)	—	with 1 pup	—	♀ with 3 pups (× 2)	—	—	—
Pups alone without mother	—	1	1	—	—	—	—
With people	14	—	—	—	—	—	—
"Greeting" group	—	—	—	—	1	1	1
Total count	175	64	33	8	10	18	35

which were foraging, two playing, and two traveling. Three early morning greeting groups of respectively 5, 6, and 7 individuals were seen. These groupings were of extremely short duration (1–3 min). Groupings of longer duration were focused around a female in heat: four groups of 3 males and 1 female, one group with 5 males and 1 female, two groups with 6 males and 1 female and three groups with 7 males around 1 female. A major stimulus for temporary pack formation in the urban dog is clearly the presence of an estrous female (see Figure 6).

Figure 6. *Free-roaming house dogs forming a temporary pack around an estrous bitch (a), creating a traffic hazard crossing highway (b), and attempting to reach the bitch who is seeking refuge under an automobile (d). The resident was unable to leave the house for fear of being attacked as the temporary pack rests on her front steps (c).*

Discussion

Against the background of Beck's (1971, 1973) studies of urban dogs, the present study focuses more specifically on the social ecology of a pack, or social unit, of three animals. Permanence of such a social unit may be rare in urban dogs, but this conclusion awaits further study. The distribution and abundance of food (i.e., wide dispersal, but in small concentrations) would favor solitary scavenging, and all other feral dogs observed during this study were seen singly and more rarely in pairs.* Groups of three or more were usually temporary packs of males following a bitch in heat. A permanent pack in the urban environment would also be conspicuous and more liable to human predation (the local dog catcher chases groups more than individuals).

In relation to the diurnal activity of other dogs, the feral trio remained undercover during the major part of the day. This may have been to avoid the heat of the day or human interactions, or both. They clearly avoided close proximity with people, while the activity of nonferal dogs closely coincided with human activity, i.e., a morning peak around 8:00 a.m. (when many are presumably let out of the house) and a peak in the early evening when dogs, people, and children would be out in the streets. Nonferal dogs were least active during the afternoons, the hottest time of the day, and, like the feral trio, were most active during the early hours of the morning.

In contrast to the regular morning visits of the trio to the park to scavenge for food and to chase squirrels, other dogs were only occasionally seen there and without any regularity in their visits. The amount of time and energy the trio spent in ranging over the park and chasing squirrels, which they were never seen to catch, was remarkable. Such unreinforced activity may be rewarding in its own right and could be interpreted as play.

The interactions between the three dogs were surprising, since much more overt communication and social interaction were anticipated. There was little "redundancy" however, and much

---

*Feral and free-roaming dogs observed in Mexico City and Madras were seen foraging singly and only occasionally in pairs (M. W. Fox).

communication was overt, involving body orientation and eye contact. It must be concluded that each member of the social unit was
fully aware of what the other dogs were doing and were intending
to do much of the time, i.e., a metacommunicative system and
context-related set of expectations were operating. In addition,
since there was no clear linear dominance hierarchy, there were
few ritualized displays of dominance and submission and no greeting rituals to the alpha individual (as exemplified by the wolf pack,
[Fox, 1971b]). Leadership, i.e., choice of direction of movement,
was usually made by the female, who at such times could be designated leader. On two occasions (while feeding) she asserted dominance over the others by staring and growling. The response was
nonsubmissive avoidance. Competitive behavior was never seen
on other occasions.

In relation to *context*, therefore, individual differences were
evident, and in different contexts the female was the leader or was
dominant. If behavior differs between individual members of a
social group in the same context, then the possibility of context-
related *roles* in relation to the social ecology of the group must be
considered. Two of the dogs of this social group appeared to have
different roles which were apparent at different times in certain
contexts. While the female was most often identifiable as the
leader or decision-maker, the male (Y) was the most aggressive
toward strange dogs and might be designated as the guard of the
unit. A specific role for the other male (X) was not so easily assigned. There was no evidence of dominance over, or subordina-
tion to, the other male, and X marked (the home range) 2.6 times as
frequently as Y (except in the park) and marked three times as
much as Y over F after she had urinated. This latter observation
may indicate a closer bond between X and F than between F and Y,
an interpretation supported further by the fact that Y left the social
unit more often than either of the other dogs. Also the female (F)
initiated more social interactions with X than with Y, while Y interacted at almost equal frequencies with X and F. There was no
evidence that the marks were either used or respected as territorial
boundaries in this feral trio.

In spite of these feral dogs' remarkable abilities to adapt to the
urban environment, it is unlikely that they could succeed in raising
a litter of pups under such conditions. Nutritional, parasite, and

disease problems, both pre- and postnatally, would reduce the chances of any offspring surviving up to weaning age. Toward the end of this study, F was reportedly in heat (September, 1973) according to one local resident, and further observations confirmed that she was pregnant (or pseudopregnant). If any pups were born, they could not have survived long since we were unable to find them although the trio was still active in the area through January. While Nesbitt (1975) has shown that there is some natural selection in feral rural dogs for a particular body size/phenotype, it is unlikely that natural selection could operate in feral urban dogs because of the high turnover rate of the population and low chance of survival for any offspring.

Summary

A group of three feral dogs (two males, one female) living in vacant buildings in St. Louis, Missouri was studied. They avoided close proximity with people and were active earlier and later in the day than the people and loose pets in the area. They found food while scavenging through human trash. The group's activities were usually initiated by the female of the group though otherwise there was no clear linear hierarchy and few ritualized displays of dominance or greeting. Specific roles within the group were observable. Though the female appeared to be pregnant during the study, puppies were never noted.

IV

Vocalizations in Wild Canids and Possible Effects of Domestication

Introduction

In order to make an inventory of the various sounds emitted by canids, the vocalizations from several different species of Canidae reported in this study were recorded, and selected sounds were analyzed on a sound spectrograph. From such an inventory, interspecies comparisons of vocalizations and of the various contexts in which they occur could lead to a closer understanding of taxonomic interrelationships. It might also throw light on the origin(s) of the domesticated dog *Canis familiaris* which is, to date, an enigma.

The study of canid vocalizations is also of significance for other reasons. First, an understanding of the sounds made by these animals is a necessary part of understanding their total communication system (Fox and Cohen, 1977), for the vocalizations not only enhance visual and olfactory displays but in some cases may serve as a substitute for these displays, as over long distances, in dense cover, or at night. Second, the canid family is a very special group in some respects. Its members come from a wide variety of social systems (Fox, 1975) ranging from the gregarious (e.g., wolves,

Table I.
CANID SOUND TYPES[a]

Social Context	Mew	Grunt	Whine (whimpers)	Yelp	Scream	Yip
Greeting	F[b]	WD[c]	WCD	D	F	C
Play-solicit	—	—	D	D	—	—
Submission	F	—	WCD	D	WCD	—
Defense	—	—	WCD	D	CFW	—
Threat[d]	—	—	—	—	—	—
Care- or contact-seeking	NF	DW	N;DWC	N of WCD; C	F	C
Distress (pain)	N	—	N; WCDF	N of WCD; D	N; WCDF	—
Contact-seeking (lone calls)	—	—	N; WCD	D	—	—
Group vocalization	—	—	WCD	—	—	C

Asiatic wild dogs) to the relatively nonsocial (e.g., foxes). Communication systems vary greatly between these two main groups (Fox, 1971b), and comparative studies may add to our knowledge of the socio-ecological adaptation and evolution of canids. This knowledge may soon prove essential to the preservation and management of endangered species.

Materials and Methods

Over a period of 7 years, recordings of vocalizations were made of the following species at various ages (the animals were all hand-raised in captivity):

Table I. continued

Yowl-howl	Coo	Growl	Cough	Bark	Click	Toothsnap	Pant
WD	F	WD	—	D	—	—	F
—	—	—	—	D	—	WD	FD
—	—	—	—	—	—	—	—
W	—	WCDF	WCFD	WD	F	WCD	—
—	—	WCDF	WCFD	WCDF	F	WCD	—
—	F	—	—	D	—	—	—
—	—	D	—	D	—	—	—
WCD	F	—	—	D	—	—	—
WCD	—	WCD	—	WCD	—	—	—

[a]Most sounds are not mixed, but some of the sounds may be mixed simultaneously or successively, e.g., whine-growl, bark-howl, yelp-bark.
[b]F, red, arctic, and grey fox; D, dog; W, wolf; C, coyote; N, neonates of these species.
[c]Also "contentment grunts" of contact in neonate W, C, and D.
[d]May serve as warning to others.

    3 wolves (*Canis lupus*) from birth to 2 years
    2 red foxes (*Vulpes vulpes*) from birth to 10 weeks
    4 red foxes from 2 weeks to 10 weeks
    4 red foxes from 2 weeks to 3 weeks
    2 grey foxes (*Urocyon cineraoargenteus*) from birth to 10 weeks
    2 grey foxes from 8 weeks to 2 years
    8 Chihuahuas from birth to 6 months
    4 Irish setter × Doberman pinscher F_1 hybrids from birth to weeks
    4 coyotes (*Canis latrans*) from birth to 3 years
    4 F_2 coyote × beagle hybrids from birth to 3 years
    2 Arctic foxes (*Alopex lagopus*) from 8 weeks to 2 years
    2 Asiatic golden jackals (*C. aureus*) from 1 week to 2 years

Recordings were made in nine different social contexts listed in Table I. In addition, recordings were also made of various zoo animals including Dingo (*Canis exfamiliaris dingo*), New Guinea singing dog (*Canis exfamiliaris hallstromi*), maned wolf (*Chrysocyon brachyurus*), culpeo (*Dusicyon culpaeus*), wolf (*Canis lupus*), coyote (*Canis latrans*), bush dog (*Speothos venaticus*), and Cape hunting dog (*Lycaon pictus*).

All recordings were made on a Sony EM-2 field recorder at 7 in. per second. South analysis was then performed with a Kay Sona-graph 6061A, with high shape (HS) and narrow band settings.

Later references to short, long, or extended sounds refer to arbitrary classes of sound duration as follows: short, 0–0.9 sec, long, 1.0–2.4 sec, extended, 2.4+ sec.

Results

CLASSIFICATION OF SOUNDS

Twelve basic canid sound types were identified in this study including whines (and longer softer whimpers), short yelps, yips, screams, barks, coughs, growls, coos, howls (or yowls), mews, grunts, and guttural clicks. Variation within these categories may occur at both the individual and interspecies levels, and sounds may vary along the dimensions of duration, frequency, intensity, cycle (rhythmicity), and context.

Not all of these basic sound types are included in the vocal repertoires of every canid species, and the same sound may be used by different species in very different contexts. Foxes, for example, are the only canids known to emit a pure scream in greeting conspecifics, and domestic dogs will bark in many situations (e.g., threat, play-solicit, contact-seeking) while foxes bark only in threat. The presence or absence of these various sounds in different canid species is shown in Table I together with the context(s) in which each sound was recorded. Further descriptions of some of these basic sound types may be found in Tembrock (1968, 1960), Bleicher (1963), Joslin (1966) and Theberge and Falls (1967).

The basic sound types listed in Table I are to be differentiated from more complex mixed sounds, which are described shortly. It should also be emphasized that under the categories of whines and coos, for example, there are clear differences between and within species in the pitch, duration, and spectrographic structure of such sounds. The focus of this study, however, is not to look at such species and individual differences within sound types but rather to make an overall survey of the occurrence of these basic sound types across species to provide a basis for evaluating the possible effects of domestication on canid vocalizations.*

The cross-species survey clearly reveals the distinction between vulpine (foxlike) and canine (dog/wolf-like) species, the former showing a high incidence of coo-calls and guttural clicks (absent in the latter), a high frequency of screams, and an absence of group vocalizations. The yip sound type was exclusive to coyotes and golden jackals, and a high incidence of barks in many different contexts was characteristic of the domestic dog. No variations in occurrence of sound types in individuals of the same species were ever recorded in different contexts, i.e., sound type and context specificity was evident for each species studied.

In addition to these vocal sound types, all species communicate to some extent by means of mechanical or unvoiced sounds such as the tooth-snapping heard in wolves, coyotes, and dogs, and panting, a play-soliciting signal of dogs and foxes (see page 135). Because these mechanical sounds are of relatively low volume and do not carry very far, they are reserved for close contact situations and are often heard during agonistic encounters.

CROSS-SPECIES COMPARISONS

Some evidence is available at this point to warrant preliminary cross-species comparisons (see Figures 1 and 2).

---

*This is not to underrate the significance of individual differences, which is in itself an important and much needed area of investigation. Many clear individual differences became evident during the studies, as for example, wolves engaged in group howling and in red and grey foxes of different sexes emitting the same coo call varying widely in pitch, form, intensity, and duration. Such variations could permit even more information exchange beyond the emotional state and intentions per se that are discussed in this chapter.

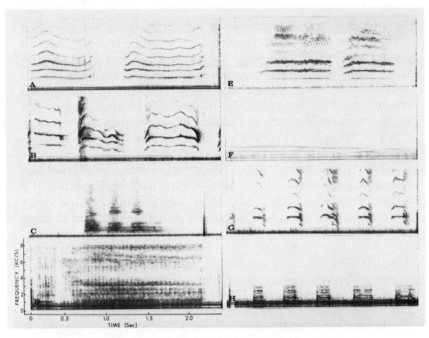

Figure 1. (A) Whines of a 1-day-old male Chihuahua. (B) Screams of a 6-day-old male Chihuahua. (C) Barks of an adult female dingo. (D) Growls of an adult male wolf. (E) Coos of a 33-day-old female red fox. (F) Excerpt from a howl of an adult female wolf. (G) Mews of a 5-week-old male Chihuahua. (H) Grunts of an 8-day-old male Irish setter × Doberman pinscher hybrid.

Whines

The whines of all species recorded were typically of short duration. Wolves (and husky dog, personal observation) will sometimes extend this sound for several seconds, however, producing what has been called an undulating whine (Crisler, 1958). This sound is created by the movement of the tongue within the vocal cavity alternately blocking and opening air passages (Crisler, 1958). The fundamental and pitch of dominant frequency (PDF) found in the whines of adult wolves were both about 1570 Hz. The same figures apply to the undulating whine.

The fundamental for the Chihuahua ranges from 400–1570 Hz, while the PDF is usually from 2000–3000 Hz. Recordings made of developing Irish setter × Doberman pinscher hybrid puppies indi-

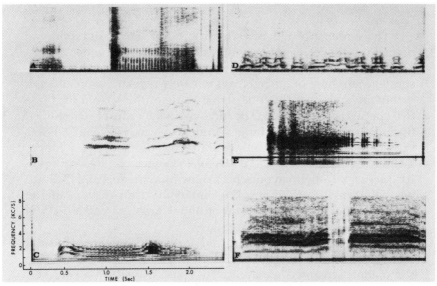

Figure 2. *Mixed sounds. Successive: (A) bark→growl of a 4-week-old male Chihuahua; (B) pure coo and coo→scream of a 33-day old female red fox; (C) yelp→growl→bark→growl of a 10-day-old male Irish setter × Doberman pinscher. Simultaneous: (D) bark-howls of a 1-year-old female wolf; (E) growl-scream of Culpeo; (F) growl-scream of a 5-month-old female Arctic fox.*

cated a fundamental and PDF of 1500–3400 Hz. Whining is most often a cyclic (i.e., rhythmic) vocalization given in distress, with the exception of the undulating whine which is noncyclic.

Yelps

These sounds, as shown by Bleicher (1963), develop in the dog in combination with the whine and later may occur separately or successively combined with a growl, bark, or whine (discussed later). This particular sound type was not recorded in other adult canids and may be a species-characteristic of *Canis familiaris*. For the purpose of this study, the yelp was regarded as a shortened, contracted form of the whine; a high amplitude piercing variant, the yip, being an analogous sound recorded in coyotes and jackals. The analogy of the yelp in the same context in the fox is the scream.

Screams

The screams of red foxes tend to be of longer duration and occur in a greater variety of contexts than those of other canids. The gray foxes recorded emitted screams which were generally shorter than those of the red foxes but were repeated more often. It is thus possible that the total signal value of the scream for these two species is about equal. The fundamental for most foxes studied (including the Arctic fox) ranged from 1200–2000 Hz. The PDF was between 2000–5000 Hz. Chihuahua recordings indicate a fundamental of 1200–2000 Hz and a PDF of 2000–3200 Hz. Recordings of Irish setter × Doberman pinschers show fundamental and PDF values between 1800–2700 Hz. Preliminary data on the screams of coyotes indicate a fundamental of about 2400 Hz and a PDF of about 2700 Hz.

The scream is noncyclic in red foxes and repeated, but not cyclically, in gray foxes. Chihuahua puppies given painful skin stimulation screamed noncyclically. Those placed on a cold surface, however, emitted cyclic screams.

Barks

Barks of all species recorded were of very short duration (i.e., 0.5 sec). All principal frequencies (fundamental and PDF) lie in the lower register between 0–2000 Hz. The main differences between the barks of various canid species concern cyclicity. The domestic dog will often bark cyclically in a singsong manner, one bark following another until a train of barks results. This is commonly heard during territorial defense and care or contact solicitation. Foxes bark noncylically, while our data indicate that wolves may bark either cyclically or noncyclically. Further investigation is needed to distinguish the different stimulus situations that elicit these two barking forms in wolves.

Growls

Growls may vary in duration from short to extended, depending on the situation and intensity of the social encounter. The growls of all species are noncyclic. While foxes growl only in threat and defense, wolves and some dogs growl while greeting one another possibly reaffirming their dominance relationships. Wolves, dogs, and especially coyotes may often growl during group vocalizations. A muffled growl-bark or cough was recorded, but not analyzed, in all canids. It may serve as a warning to off-spring and others in agonistic contexts.

Howls

All recorded howls of wolves and coyotes were of long to extended duration. The fundamental for both species was between 400–2000 Hz and the PDF from 1200–2900 Hz.

Mews

Preliminary evidence indicates that newborn to 5-week-old red and grey foxes tend to repeat their neonatal mews more often than dogs of the same age group. Foxes maintain this sound as part of their vocal repertoires throughout life, while it is only heard in neonates of all other canids studied.

Grunts*

The fundamental of the grunts of wolves and dogs ranged from 85–200 Hz when detectable. The sounds are heard in neonates of these two species and the coyote but were not recorded in foxes.

Coos

This sound is heard only in foxes. One form is trill-like, while the other is more of a cackle. Differentiation between the two forms on the basis of spectrographic evidence is difficult, and it is not yet clear which stimuli elicit each form of the coo.

MIXED SOUNDS

The above sound types may be mixed in one or two ways: either by superimposition of two vocal sounds or by successive emission of two or more types of sounds (vocal and/or mechanical) (see Figure 2). A combination of these two mixed forms is also possible, so that a mixed sound, for example, might successively follow a pure sound. The phenomenon of sound mixing correlates well with the superimposition of body language postures and facial expressions (Fox, 1971b). A wolf faced with a situation simultaneously eliciting

---

*This sound type is to be distinguished from the groans and grunts of a sick or dying canid, or of one in intense pain as during parturition.

Table II.
MIXED SOUNDS BY SUCCESSIVE EMISSION OF TWO[a] SOUND TYPES

First sound emission	Second sound emission										
	Whine	Scream	Bark	Growl	Coo	Howl	Mew	Grunt	Pant	Cough	Click
Whine[b]	—	—	D[c]	WC	—	CW[d]	—	—	D	—	—
Scream	—	—	—	DC	—	—	—	—	F	—	F
Bark	D	—	—	D	F	DWC	—	—	—	—	—
Growl	WDC	—	WC DF	—	—	—	—	—	—	F	F
Coo	—	F	F	—	—	—	—	—	—	—	—
Howl	W[d]	F	—	—	—	—	—	—	—	—	—
Mew	—	—	—	—	—	—	—	WD (cubs)	—	—	—
Grunt	—	—	—	—	—	—	WD (cubs)	—	—	—	—
Pant	D	F	—	—	—	—	F	—	—	—	—
Cough	—	—	F	F	—	—	—	—	—	—	F
Click	—	F	F	F	—	—	—	—	—	F	—

[a]Complex combinations of more than two sounds are also possible, e.g., cough → click → growl-scream

[b]Includes abbreviated form—the yelp and yip.

[c]D, domestic dog; w, wolf; c, coyote; f, red fox.

[d]Howl preceded by whining in wolf and yipping in coyote but not strictly mixed.

Table III.
MIXED SOUNDS BY SIMULTANEOUS EMISSION OF TWO SOUND TYPES

	Whine	Scream	Bark	Growl	Coo	Howl	Mew	Grunt	Pant	Cough	Click
Whine[a]	—	—	D[b]	WDC	—	—	—	W	C	—	—
Scream	—	—	DF	FCW	F	—	—	—	—	—	—
Bark	—	—	—	WD	F	W	—	—	—	—	—
Growl	—	—	—	—	—	—	—	—	—	—	F
Coo	—	—	—	—	—	—	—	—	—	—	—
Howl	—	—	—	—	—	—	—	—	—	—	—
Mew	—	—	—	—	—	—	—	—	—	—	F
Grunt	—	—	—	—	—	—	—	—	—	—	—
Pant	—	—	—	—	—	—	—	—	—	—	—
Cough	—	—	—	—	—	—	—	—	—	—	—
Click	—	—	—	—	—	—	—	—	—	—	—

[a] Includes abbreviated form—the yelp and yip.
[b] D, domestic dog; W, wolf; C, coyote; F, red fox.

fear and aggression may concurrently display facial and body characteristics expressive of both motivations (Schenkel, 1947). The accompanying vocalizations are likely to include components that may be heard separately in pure form in fearful and aggressive contexts.

Tables II and III indicate the mixed sounds that have been positively identified, and it is likely that further investigation will reveal more sound combinations.* (See also Figure 2.)

Less social canids such as foxes do not mix sounds to as great an extent as do the more gregarious canids (see Table II and III). Foxes tend to live solitary lives and coming together briefly during the mating season; thus much of their vocal communication relies on calls that must be audible over long distances in order to be effective. Since these animals spend relatively little time in social units, a refined, graded system of low intensity vocalizations would have low selection priority. Gregarious canids such as the wolf and dog, however, tend to mix sounds more than the solitary canids, and facial and body expressions are also more highly differentiated (Fox, 1975). Thus, with increasing social complexity in the Canidae, there is also an increasing complexity of the communication repertoire.

SOUND DEVELOPMENT

We have recorded developmental sequences for two domestic dog breeds: Chihuahua and Irish setter × Doberman pinscher. Data on the sound development of the red fox are available in Tembrock (1958).

Chihuahua
The Chihuahua can whine, scream, grunt, and mew at birth. All of these sounds are functionally similar as they serve to alert the mother to the distress and location of her pups. Without the pres-

---

*Two mixed sound types were recorded but not analyzed spectrographically in domestic dogs. One is baying, a combined bark-howl characteristic of tracker dogs when on the scent. The other, yowling, is a combined and repeated yelp-bark and is possibly analogous to the group howling of wolves heard as an early morning chorus among neighborhood dogs.

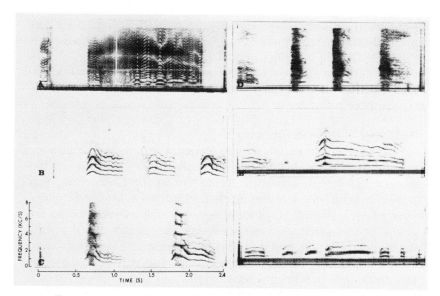

Figure 3. (A) Complex long grunt of a 1-day-old female
Chihuahua. (B) Yelp of a 10-day-old male Chihuahua. (C) Transitional bark-yelp of a 2-week-old male Chihuahua. (D) Bark of a
5-week-old male Chihuahua. (E) Sharp yelp with some vertical
stratification (final segment) of a 1-day-old male Irish setter ×
Doberman pinscher. (F) Whine with coo component of a 26-day-old
male Irish setter × Doberman pinscher.

ence of these sounds at birth, survival of the pups could be endangered.

At 1 day of age, the Chihuahua may make a sound which is
best described as an extended grunt. Spectrographically, however,
the sound is more complex than the grunt (Figure 3A). It is wide-
banded with several frequency variations and occurs when the
animal is handled.

By day 6 the pup's vocal capacity increases and frequency
variations increase. At day 10 sharp frequency rises may occur,
sounding most like a yelp. The spectrograph indicates that the duration and frequency range of the rise itself approximates those of the
fully developed bark. Thus, it seems that this yelping sound may
be a developmental precursor of the bark. Further evidence for this
comes from an intermediate stage between the yelp and the bark
which can be heard (or better, seen spectrographically) at 2 weeks of

age, when differentiation and mixing of sounds begins to take place. By about the 4th week of age, the development of the bark is complete (see Figure 3B, C, and D).

At day 18 a clear successive emission of a mixed bark-growl may be heard. By the 4th week this combination of sounds becomes more refined.

Irish setter × Doberman pinscher

As in the Chihuahua, whines, screams, grunts, and mews are present at birth as distinct sound forms. At day 1 the sharp frequency rise (yelp), appearing at day 10 in the Chihuahua, was noted. In addition, some vertical stratification is visible from the spectrograph (Figure 3E), indicating a possible relationship to the growl. By day 5 frequency variations become greater. The first fully developed bark was noted on day 10. Mixing also begins to occur widely on this day as successive emission of sounds begins. Mixing by superimposition begins between the 2nd and 3rd week of age. On day 26 a call was recorded when a pup was isolated. Spectrographic analysis later identified the call as having the short vertical frequency changes typical of the coo call of foxes. Cooing, however, is not heard in dogs in its pure form, and this recording may provide further evidence for the evolutionary link between foxes and domestic dogs (see Figure 3F).

Developmental data indicated that domestic dogs begin to mix sounds first by successive sound emissions and later by superimposition. The Chihuahua seems to develop its vocal repertoire somewhat more slowly than the mixed breed of Irish setter × Doberman pinscher. It should be noted that Doberman pinschers have been specifically bred as guard dogs, and the sooner they begin to vocalize the sooner they may be put to use. In addition, the Chihuahua has been bred for neoteny, a state of somewhat retarded development, and its schedule of vocal ontogeny may thus have been affected to some degree by the domestication process.

Discussion

The 12 sound types presented here represent an attempt to classify the major components of a complex vocal communication system in Canidae.

Developmental data indicate that the earliest sounds present in canids are those which elicit approach of conspecifics, specifically the mother. These sounds serve to decrease social distance. Once this distance is decreased, the mother may become aware of the needs of her young and may then serve these needs. On several occasions while recording the sounds of neonate domestic dogs, the mother was visually isolated from her pups but well within hearing range. As soon as the pups began to vocalize their distress to various experimental conditions, the mother made vigorous attempts to physically contact her young. Thus it seems that the distress calls (e.g., whines, screams) are releasers for some maternal behaviors, particularly retrieval. The way in which the releasing mechanism works is unclear, but it may be possible that the mother is (hormonally) predisposed (or sensitized) to interpret these sounds as noxious/aversive stimuli and acts in such a way as to cease them.

A similar interpretation may be applied to some agonistic vocalizations. Captive coyotes have been observed emitting long screams when threatened at a distance by dominant animals who had previously attacked them. The defensive scream often served to cut off the attack. Here, again, the sound may be aversive or noxious stimulus to the dominant animal.

The second set of sounds to develop are those which may elicit withdrawal of conspecifics increasing social distance. These sounds (e.g., barks, growls, clicks) are not present at birth but develop between 1 and 4 weeks of age. At this time the animals are becoming more self-sufficient and are beginning to explore their social environments. It is at this stage, then, that distance-increasing signals become as necessary as those which decrease social distance.

If we broadly classify canid vocalizations as those eliciting approach or withdrawal (A-type or W-type, respectively; see Schneirla, 1959), two separate continua of sound progression be-

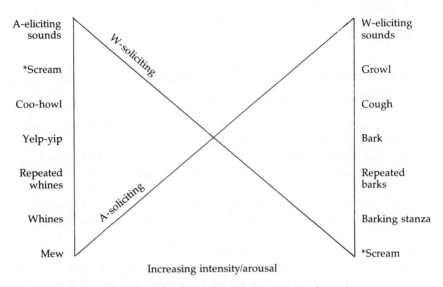

Figure 4. *Hypothetical relationships between approach- and withdrawl-eliciting sounds and increasing intensity/arousal. *Paradoxically, screams, with intense arousal, may be noxious (W-eliciting) or attention/care-soliciting (A-eliciting).*

come evident (see Figure 4). Considering first the A-types, the mew is offered as the simplest expression of a motivation/need for the approach of a conspecific. The more intense versions of this sound are respectively the whine, repeated whine, yelp, and contact scream (of foxes). All of these sounds are related functionally, and the relation of specific physical sound features may be seen spectrographically.

The growl (and nasal variant, the snort) seems to be the least intense vocal expression of a motivation/need to increase social distance and is thus the starting point of the second continuum (W-type).† Increasingly intense expressions of the same message may take the form of growl-barks, barks, randomly repeated barks, and uniform barking stanzas, particularly in the domestic dog.

---

†Repeated, low and extended growls may also occur rarely during greeting and while being petted in some breeds of domestic dog, notably in malemutes and huskies. These sounds, possibly a modified form of contentment grunt, may be analogous to purring in felines.

Successive bark-howls and short rolling howls comparable to a barking stanza associated with defensive aggression in the wolf have been recorded. This is interpreted not as an evolutionary relationship between barking and howling but rather as a species typical phenomenon, where howling in the wolf (like barking in the domestic dog) may be at a low threshold and be evoked in a variety of different contexts.

In addition to the A-eliciting and W-eliciting sound classes, Tembrock (1968) suggests two further sound groups: warning sounds and infantile sounds. The warning sounds (e.g., bark) serve as an alarm call for the entire social group. Rather than increasing social distance between members of the group (as with W-type sounds), these sounds tend (potentially or actually) to increase the entire group's distance from an external danger. These sounds often occur in the defense of group territory or when one animal alerts its group members to a nearby predator.

The infantile sounds are those which have become derived and emancipated and reoccur in adulthood. The mewing sound of adult foxes is one possible example of such a derived vocalization, as is the yelping of adult dogs (see later). The former vocalization occurs first in the neonate and serves to release maternal caregiving behavior and later reappears during courtship and mating.

Figure 4 further illustrates the relationship of A-eliciting and W-eliciting sounds when viewed with respect to the animal's level of motivation. As motivation for approach increases, more intense A-type sounds may be emitted. Similarly, as motivation for withdrawal increases, more intense W-type sounds may be emitted. Certain sounds, however, cross over these basic relationships and may be heard in either A-eliciting or W-eliciting situations; these sounds occur at the higher levels of motivation. Screaming, for example, may occur either as an intense contact call or as a defensive threat vocalization in foxes. An analogous situation in humans might be the observation that when highly aroused, sounds such as laughing or crying may cross over and occur in apparently inappropriate contexts (Darwin, 1873).

The group howl in wolves can have paradoxical consequences. It is clearly an A-eliciting sound, bringing animals close together for its duration and being preceded by whining and active submission—a coming together greeting ritual (Fox, 1971b). On certain occasions, however, the group howl may be aborted and aggressive behavior may be released. This has been observed upon numerous occasions in captive wolf packs and also among captive groups of coyotes. A tape-recorded howl or a siren may be used to evoke group howling, but at the point of coming together, any conflicts present between group members may be reactivated as a result of the ensuing close proximity.

A particular facial expression, open mouth "play face," coupled with panting in the domestic dog may be analogous to laughter in man (Fox, 1971b) occurring in similar contexts, namely greeting and play-soliciting. In the red fox, panting may be accompanied by muffled screams, mews, and purring sounds during greeting. The excited panting, a metacommunicative play signal, exemplifies how emotional changes may influence the rate and depth of respiration.

Our data, particularly that concerning mixed sounds, imply that canid vocal communication is essentially an emotional language, comparable to the intentionality expressed in nonvocal body postures and facial expressions. This is supported by the fact that continuous gradations of sound types are possible indicating the degree of arousal motivation and intentions in a given context. It is analogous therefore to the "paralanguage" or emotional overtones of human speech. This would correspond to mixed or ambivalent motivational states. Experiments by Tembrock (1958) indicate that the rhythm of the sounds of cubs separated from their mother is more important in releasing the mother's retrieval behavior than the phonetic characteristics of the sounds themselves. This, too, lends support to the idea that it is not necessarily so much what is said that is of significance but rather how it is said.

In the course of experimentation with rearing domestic dogs in social isolation (Fox, 1971a), it was found that such animals, although capable of producing all normal sounds, vocalized much less frequently than those raised with conspecifics. When introduced at a later age to other animals, however, these isolates began to vocalize at a level comparable to those raised in social groups. Winter et al. (1973) report the same phenomenon in squirrel monkeys (Saimiri sciureus) which had been deafened at birth or auditorially isolated. Their subjects also increased their frequencies of vocalizing when introduced to conspecifics. These observations indicate that some type of social reinforcement or facilitation is involved in achieving full vocalizing potential.

The fact that squirrel monkeys deafened at birth are capable of producing all normal vocalizations indicates that the vocal repertoire is largely genetically predetermined. This seems to be the case in domestic dogs as well since experiments by Fox (1971a) indicate that no auditorially evoked potentials could be recorded from the dog's brain until 2 weeks of age. In effect, the dog is deaf until this age; but vocalizations begin to develop at (or possibly before) birth, and the basic sound repertoire is complete by about 4 weeks of age.

In addition, hand-raised domestic dogs fed and handled on a strict schedule quickly decreased their frequencies of care-soliciting distress vocalizations, presumably because their sounds were in no way positively reinforced by petting or feeding (Fox, 1971a).

When placed in unfamiliar surroundings, domestic dogs generally directed contact-seeking sounds toward the experimenter, often barking, whining, and yelping, while wild canids tended to be silent. This supports other evidence that there may be a genetic predisposition toward dependency in the dog which is not present in wild canids (Fox, 1971b).

Barking is another sound form which is of unusual interest as it seems to have hypertrophied in the domestic dog and is now used in a much greater variety of contexts than among wild canids. As Table I indicates, dogs may bark during greeting, play-soliciting, threat, defense, care-soliciting, distress, contact-seeking, or during group vocalizations. Barks may be simple or complex, e.g., growl-barks, repeated barks with howl-like endings and yelp-barks. This contextual variety indicates that the sound itself may not always convey specific information but rather attracts the attention of the receiver. The more specific information to follow would then be received through other sensory channels— visual and/or olfactory. These cues would then identify the meaning of the accompanying barks.

From the inventory of various sound types, a clear distinction can be made between foxes and other canids, the vulpine and canine subgroups of the genus *Canis* being evident. The similarities between wolf, dog, and coyote vocalizations point to taxonomic affinities but add little to our understanding of the origin(s) of the dog prior to domestication. The outstanding feature of the domestic dog—barking—may be attributed to artificial selection. A good house dog barks at intruders. Barking in all wild canids similarly serves as a warning or threat to intruders (except during group vocalization) and is context specific. Is the lack of context specificity in the dog a hypertrophy of domestication, or does it point to a very different nonwolf canid ancestry? Surely it would be of little advantage (to man) to selectively breed dogs that bark with virtually no context-related specificity, although it would be most difficult to exercise context-specific selection. An alternative view is that the selection pressure for silence necessary in a wild predator has been relaxed. This, coupled with a high degree of dependency and varying degrees of neotenization or infantilism with care-soliciting behavior and vocalizations in adulthood directed toward the owner, would suggest that both domestication and socialization early in life have a profound influence on the dog's vocal repertoire. The origin(s) of the dog therefore still remains an enigma although one might conclude on the basis of this study that if the wolf were the sole progenitor of the dog, then dogs would howl more and bark much less than they do. Dogs also yelp in many different contexts compared to other canids. The prevalence of this sound may indicate a clear taxonomic difference between species. Another acceptable interpretation is that the yelp, an infantile sound, persists into maturity and occurs frequently in distress/attention-seeking contexts in dogs because of the dependency promoting neoteny or infantilism induced through domestication (since the more dependent the dog is, the more trainable it is) (Fox, 1972).

Summary

On the basis of spectrographic evidence, it has been possible to identify 12 basic vocal sound types of canid species.

Vocalizations may be mixed by successive emission of two or more sound types, by superimposition of these sounds, or by a combination of these two forms. The same basic sound type may differ among canid species along the dimensions of sound duration, separation time between consecutive sounds, principal frequencies, cyclicity, and context.

Developmental data indicate that domestic dogs first begin to mix sounds by successive sound emissions at about 10 days of age and later by superimposition between 2 and 3 weeks of age. The frequency of occurrence of the basic sound types in different contexts varied between species but not within species. The possible effects of domestication on canid vocalizations are discussed.

V
Behavior Genetics of F₁ and F₂ Coyote × Dog Hybrids

Introduction

This study represents an interim review of the behavior and morphological features of F_1 and F_2 coyote × beagle hybrids. Earlier studies by Silver and Silver (1969) and Mengel (1971) are lacking in detailed analysis of the behavior of parent stock, notably of the fixed action patterns which to the ethologist can have as much taxonomic value as morphological features to the comparative anatomist. Several species-typical action patterns have been identified for the coyote (*Canis latrans*) and for the beagle (*Canis familiaris*) in our laboratory and quantified ontogenetically in several animals (Fox, 1969a and b, 1970, 1971a and b; Fox and Clark, 1971; Bekoff, unpublished observations). The purpose of this section is to present an integrated overview and to discusss some intriguing findings which have not as yet been considered.

Materials and Methods

Developmental data were obtained from 4 male and 4 female beagles, 6 female and 4 male coyotes, and 16 F_2 coyote × beagle hybrids, 6 of which were females; of this group, which comprised six litters, 2 males died prior to weaning. Two male and two female F_1 hybrids were obtained as adults and no ontogenetic data were available for these animals which were bred by artificial insemination (courtesy of Dr. J. J. Kennelley, Bureau of Wildlife and Fisheries, Denver, Colorado). All subjects studied developmentally were hand-raised by stomach tube on Esbilac®, a synthetic bitch's milk formula (Borden Company). The various morphological and behavioral characteristics listed in Tables I and II were looked for. Much of the behavioral data were collected from dyads interacting between 21–50 days of age and subsequently from groups of animals housed in groups of three or four. The most significant findings to date are reviewed (see Figure 1).

Results

MORPHOLOGICAL CHARACTERISTICS

All F_1 hybrids had flop or pendulous ears, bushy coyotelike tails with an active supracaudal gland. Dewclaws were absent. The body type of the F_1's is best described as mesomorphic, being stockier than the coyote, but lighter and leggier than the beagle. All F_1 hybrids had a smooth coat with guard hairs slightly longer than the beagle but shorter than the coyote; the winter underhair was much thicker than in the beagle. (See Table I and Figure 1.) It should be added that two of the F_2 canids, both with piebald coloration, had a medium (coyote) length coat that was broken rather than smooth. The remaining 12 F_2 hybrids had coats intermediate in length and with underfur density between coyote and beagle, but all coats were smooth like the beagle (Table I). Their coat color varied, being dark sable in the winter and paler in the summer.

The F_2 generation showed independent segregation of beagle and coyote characteristics (see Table I). Erect ears, a coyote characteristic, were less frequently seen than pendulous or semipendulous ears. All animals had fairly bushy tails and two had dewclaws on the hind legs. The piebald coat color (white and brown/black) of the beagles was seen in a low percentage of F_2's, most of the ani-

Table I.
FREQUENCIES OF MORPHOLOGICAL CHARACTERISTICS OF BEAGLES, COYOTES, F_1 AND F_2 HYBRIDS

Animal	*Characteristics*											
	Number of subjects	Erect ears	Semi-erect ears	Flop ears	Bushy tail	Active tail gland	Dewclaws (hind legs)	Dark neonate coat color	Piebald color	Wild type	Morpho-type	Eruption of upper temporary canines
Coyote	10	10	0	0	10	10	0	10	0	10	Ecto/meso-morph 10	12-14 days
Beagle	8	0	0	8	0	0	4	0	8	0	Meso/endo-morph 8	20-24 days
F_1 Hybrid	4	0	0	4	4	4	0	Not observed	0	4[a]	Mesomorph 4	Not observed
F_2 Hybrid	14	2	4	8	12	8	2	12	2	12[a]	Ecto/endo-morph 8/2[b] (4 mesomorph)	17-22 days

[a]Seasonal variations from pale to dark coat color.
[b]Figures denote numbers of animals.

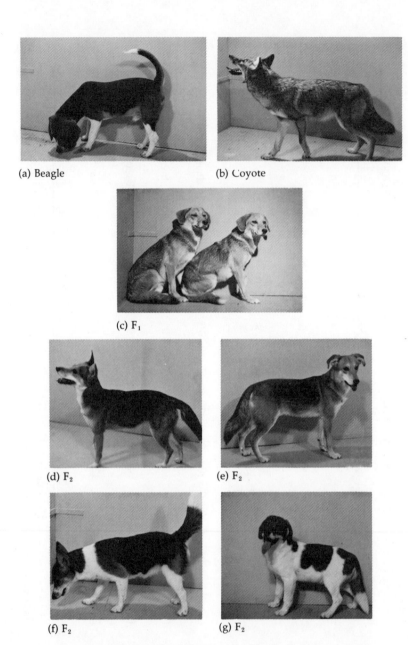

(a) Beagle

(b) Coyote

(c) F_1

(d) F_2

(e) F_2

(f) F_2

(g) F_2

Figure 1. *Phenotypes of beagle (a), coyote (b), F_1 (c), and F_2 (d,e, f,g) hybrids. Note independent segregation in F_2's—(d) ectomorph; (e) ecto/mesomorph; (f) endomorph; and (g) mesomorph.*

mals being either pale or dark sable without any large areas of white on the body, although most F_2's had white around the muzzle and lower jaw (i.e., white facial mask of the coyote) and also on the tail tip. The dark sable type tended to develop a pale coat in the summer months like the F_1 hybrids (see Figure 2). The most frequent somatotype was a long-legged coyotelike ectomorph, although a smaller percentage of endomorphs, which were stockier and sometimes shorter-legged than the beagles, were seen. The tail gland odor was harder to detect and seemed less active in the beaglelike F_2's compared to their ectomorph littermates. These endomorph-type hybrids also had an overwhelming "doggy" smell, much stronger even than in beagles.

Eruption of the upper temporary canine teeth in the F_2 canids was usually earlier than in the beagles, the age of eruption being extremely early in the coyote compared to the domestic dog (Table 1).

One of the most intriguing morphological characteristics was the neonatal coat color, the significance of which is discussed subsequently. All wild canids are dark at birth (Fox, 1971b) and do not possess the adult coat color, while most domestic dogs show the adult color at birth. The majority of the F_2 hybrids had a dark dusty-brown neonatal color, which eventually gave way to a pale or dark sable. Two animals (Table I) had the adult color at birth, namely the beaglelike piebald coloring.

BEHAVIORAL CHARACTERISTICS

Both the F_1 and F_2 hybrids were timid and shy of strangers. The endomorphic beaglelike F_2's were less timid than the more ectomorphic coyotelike F_2's. (See Figures 3 and 4.) Coyotes, even after being hand-reared, become increasingly shy of strangers with increasing maturity (Fox, 1971b). The wildness of these canids, or capacity to regress to a feral state, was accidentally evaluated at our field station. Two F_2 hybrids and one beagle escaped from their enclosure; the latter animal stayed close to the enclosure and was easily caught, while the hybrids, both ectomorphic or coyote types, roamed the 2000-acre station for several days before hunger brought them back and they could be coaxed back into the pen

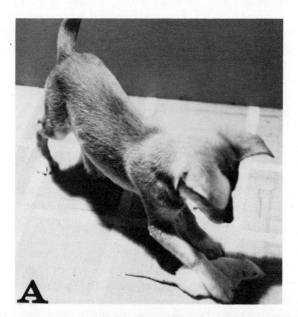

Figure 2. *Coat color change with maturity in* F_2 *hybrid. (A) Pale sable at 12 weeks (note coyotelike prey-catching forelimb stab); (B) same animal at 1 year with dark sable coat.* Facing page: *(C) phenotypic variation in* F_3 *hybrids (note piebalding and wide variation in ear positions).*

with food. Three similarly raised coyotes of the same age were released at the field station and were provided with food at the spot where they were released. They moved away permanently from this site after 4 days, and were not seen again.

As in wild canids, the first signs of shyness toward novel stimuli (environmental fear) began to develop around 5 months of age in a high percentage of the F_2 hybrids. The wild temperament trait is discussed in detail earlier (Fox, 1971b).

Both coyotes and F_1 hybrids showed marked aggression toward members of the same sex (Table II). F_2 hybrids also showed increasing proximity intolerance for members of the same sex, although unlike the coyotes (Fox and Clark, 1971), they did not fight seriously between 4 and 5 weeks of age. Dominance fights occurred later, around 6–9 weeks, and permanent hierarchies were formed shortly after this age. By 6–8 months, intrasexual aggression was marked and was the cause of death of one subordinate male F_2 hybrid that for necessity had to be housed with male and female littermates.

Clasping a conspecific around the waist during aggressive interactions (Fox, 1969b), a typical action pattern occasionally observed in the beagle, was frequently seen in both F_1 and F_2 hybrids.

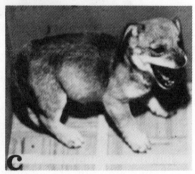

Figure 3. *Agonistic interactions in coyotes and* F_2 *coyote* ×
beagles. (a) Intense intraspecific aggression and high proximity in-
tolerance in 29-day-old coyotes; (b and e) defensive threat gape in
coyote and hybrid; (c and d) more submissive-defensive gape in F_2
hybrid, similar to coyote (f). See facing page.

Aggressive displays in the coyote include horizontal extension of
the tail, while in the beagle it is arched in a more vertical position
(Table II). In the F_1 hybrids, both tail positions were seen, and often
a *compromise* tail position, between the vertical and horizontal, was
observed during agonistic encounters. F_2 hybrids showed both
coyote, beagle, and intermediate type tail positions, the vertical
position occurring in the beaglelike endomorphs and the horizon-

tal position in the ectomorphic types. The latter hybrids more often showed the intermediate position than the former, and the tail at rest was held down as in the coyote.

Threat-gapes (Fox, 1970) and hip-slams (Fox, 1969b) are action patterns common to the coyote during agonistic interactions and have not been observed in beagles. Both actions were seen occasionally in the F_1 hybrids at lower amplitude and frequency; F_2 hybrids showed a greater range of variation in occurrence of the threat-gape (Table II), but all showed the hip-slam during fighting and play-fighting.

The prey-securing forelimb stab and ability to catch, kill, dissect, and ingest live prey (4-week-old rats) will be quantifed in detail in Chapter 6. Most F_2 hybrids exhibited the forelimb stab of the wild canid, and the majority of these animals were efficient at killing and ingesting prey. Those that failed (interestingly, the

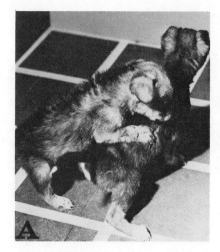

Figure 4. *Action patterns typical of coyote in* F$_2$ *coyote* × *dog hybrids: (A) standing-over, (B) threat-gape, (C) inguinal response (hind leg is raised).*

beagletype endomorphs) in executing a normal coyote-killing sequence showed inhibition of the prey-killing bite, so that injured prey were eaten alive. In coyotes, no bite inhibition was observed, while in beagles, bite inhibition, plus inhibition of eating, were recorded.

Table II.
OCCURRENCE OF OBSERVED BEHAVIORS IN COYOTES, BEAGLES, F₁ AND F₂ HYBRIDS

Animal	Number of subjects	Shy temperament	Intrasexual aggression	Aggressive clasping	Vertical aggressive tail	Horizontal aggressive tail	Threat-gape	Hip-slam	Forelimb stab	Prey-killing at 8 weeks	Inguinal response	Estrus
Coyote	10	8	8	10	0	10	10	10	10	10	10	Early spring (February)
Beagle	8	0	6	4	8	0	0	0	2	0	1	Spring and fall
F₁ Hybrid	4	4	4	4	4	4	2	4	Not tested	Not tested	4	Late fall–early winter (October–November)
F₂ Hybrid	14	8	12	14	6	8	14	14	14	10	14	Winter (December) occasionally spring and fall

The inguinal response, a display of considerable ontogenetic and social significance in Canidae (Fox, 1971d), occurs rarely in beagles but consistently during social interaction in coyotes. This display was seen in both F_1 and F_2 hybrids (Table II).

During friendly social approach, the more social canids, such as dogs and wolves, show a licking intention, often combined with a submissive "grin" (Fox, 1970). In coyotes, this licking intention is rarely seen and at low frequency. A slightly higher frequency was noted in the F_1 hybrids and in some individual F_2's; in the latter there was no apparent correlation with the somatotype.

Estrus in the coyote occurs in early spring (February) so that the birth of cubs is timed with a relative abundance of prey. In the beagle, there are two heat cycles per year, usually in the spring and fall (but in many individuals at almost any month of the year). Estrus in the F_1 hybrids was restricted to late fall–early winter. The F_2 hybrids were more variable, estrus occurring at any time from late fall to spring; only one heat was recorded during the summer. One F_2 animal had two heats (spring and late fall), but no fertile mating occurred.

As early as 3 weeks of age, a marked difference in response to handling was seen in F_2 hybrids. Some pups felt more tense and had a greater muscle tone characteristic of infant coyotes, while others remained relaxed, almost flaccid, when picked up, like beagle pups. There was no correlation with these two types and their somatotype. The more tense pups were more easily alarmed by sudden noises and would invariably give an open-mouth threat-gape which is first seen at this age and similarly evoked in coyotes (Fox, 1970). An effective, although unsophisticated, test for pain-induced aggression first used in studying individual differences in coyote cubs was used on these hybrids. The test consisted of holding the cub in one hand, abdomen uppermost, and with the other hand alternately and mildly pinching the cheek skin. In the relaxed hybrids, there was either distress vocalization or no response. The tense hybrids reacted like coyote cubs with defensive threat-gapes and at a later age (4 weeks) with growls and vertical retraction of the lips. Repeated stimulation (10–15 times) would produce a ragelike reaction and self-directed aggression where the subject would seize and bite one or both of its forepaws. This had a vicious circle effect, self-inflicted pain increasing the aggression. Movies taken of these two different sets of reactions dramatically illustrate this phenomenon, which may be attributed to threshold differences in responsiveness to mild cutaneous stimulation. One of the tense hybrids at 4 weeks of age had to be forcibly restrained and the canine teeth clipped, after a brief fight with a conspecific which triggered self-directed biting; it was unable to walk for several days, both forelimbs being severely mutilated. No such extreme reactions (low threshold?) have been observed in coyote cubs.

The pups identified as tense individuals at 3–4 weeks were generally more efficient in handling live prey, showed less proximity tolerance and greater intrasexual aggression, and were actually dominant over their relaxed conspecifics. Future studies will attempt to quantify such early individual differences in responsiveness to cutaneous stimulation.

ADDITIONAL OBSERVATIONS

The majority of F_2 hybrids had a characteristic defect in that the tip of the tongue often protruded out of the mouth while at rest. Measurements have not been taken, but this may be either a neurological defect or a disproportion between tongue and muzzle length. When alert, all but 2 of the 14 F_2 hybrids could completely retract the tongue. This anomaly was not seen in the F_1 hybrids.

The endomorphic F_2 hybrids also showed varying degrees of prognathism, the lower jaw being shortened. Dentition appeared normal and in no instance was the malocclusion severe enough to interfere with the alignment of the permanent teeth.

All endomorph beagletype F_2 hybrids had a delayed closure of the parietal fontanelle (4–6 weeks). Like beagles, these latter animals were more vocal than the ectomorph type. The F_2's had similar vocalizations, namely a bark-howl which, as in the F_1 hybrids, sounded like a successively combined beagle and coyote call.

Discussion

The morphological and behavioral data strongly suggest that the following traits have a dominant mode of inheritance: bushy tail, pendulous ears, sable wild color, short guard hairs and smooth coat, seasonal change in coat color, active supracaudal gland, dark neonatal coat color, coyote white facial mask (around lower jaw and muzzle), coyote ectomorph somatotype, one annual estrus, intrasexual aggressivity, aggressive clasping, hip-slam, threat-gape, inguinal response.

The dark neonatal coat color, a characteristic of all wild canid species so far studied (Fox, 1971b), may be related to heat conserva-tion since neonates are partially poikilothermic. Most domestic breeds of dog have the same coat color at birth that they will have as adults, though a greater number show a seasonal change in coat color.

Eruption of the upper temporary canine teeth, which is very early in wild canids compared to all domestic species so far studied, was earlier in both F_1 and F_2 hybrids. As in the beagle, erup-tion of the lower canines occurred later in the F_2 hybrids, and the upper incisors erupted earlier than the lower, the central pair usu-ally being the last to emerge. As would be predicted on the basis of independent segregation of the F_2's into beagle meso/endomorphic and coyote (ectomorphic) types, tooth eruption was earlier in the coyote type.

Mengel (1971) also noted a greater variation in F_2 hybrids from doglike to coyotelike compared to the more intermediate F_1 hybrids. However, he did not report any detailed developmental data, but it is significant that all his hybrids were aggressive and fought a great deal early in life. A similar increase in aggression, compared to the beagle, was apparent in all F_1 and F_2 hybrids in the present study.

Mengel reported small litters from the F_1 hybrids, and this corroborates the present findings where the average litter size was 2.7. This may be indicative of a lower fertility in these hybrids.

Gier (1975) and Mengel (1971) have both remarked that shift in timing of estrus in coyote × dog hybrids would prejudice survival of any offspring born in the wild during the winter (except in temperate southern regions of the United States). These data add further weight to their argument. Kennelly and Roberts (1969), in a study of the F_1 hybrids used in this study, found evidence for seasonal spermatogenesis in the males.

The inguinal response, a highly ritualized social display in coyotes which has a clear ontogenetic history (Fox, (1971b), was also seen at high frequency in the F_2 hybrids but only rarely in beagles. Its function early in life may be to truncate agonistic interactions prior to the establishment of social relationships. The high frequency of this display correlates with the great incidence of fighting during the first 3–5 weeks of life in both coyotes and F_2 hybrids compared with the beagle.

Behavioral and developmental studies of these canid hybrids have made it possible to identify a number of behaviors which, as emphasized earlier, may be as valuable as morphological traits in facilitating the identification of patterns of inheritance and of possible ancestry. Of particular interest and a focus for future research on subsequent generations and back-crosses is the ontogeny of agonistic behaviors and quantitative analysis of various displays and action patterns. The canid family holds a wealth of possibilities for further studies of behavior genetics since many species (wolves, dogs, coyotes, and jackals) (Gray, 1954; Kolenosky, 1971) will hybridize, and for each of these species, characteristic action patterns and displays have been identified and detailed ethograms and developmental data are available (Fox, 1971b).

Summary

The frequency of occurrence of various morphological and behavioral characteristics are documented in coyotes, beagles, and their F_1 and F_2 hybrids. These data reveal certain consistencies in the patterns of inheritance of such beagle traits as pendulous ears, smooth, short coat; and coyote traits, such as ectomorphy, sable coat color, dark neonatal coat color, bushy tail with active supracaudal gland, and white facial mask. The above traits were most frequently seen in the F_2 hybrids, while the F_1 hybrids tended to be intermediate between the dog and coyote and extremely uniform in appearance. In the F_2 hybrids, independent segregation was apparent, most offspring being coyotelike ectomorphs with pendulous or semipendulous ears, while others were beaglelike endomorphs, but with proportionately shorter legs and longer bodies than actual beagles. Both F_1 and F_2 hybrids resembled the coyote in terms of shyness, intrasexual aggression, and occurrence of several action patterns associated with aggression and social interaction.

VI

Effects of Domestication on Prey-Catching and Killing in Domestic and Wild Canids and F_2 Hybrids

Introduction

Following earlier developmental studies of prey-catching and killing behavior in various species of Canidae (Fox, (1969a), the present investigation focuses upon differences in organization and temporal ordering of these behavior patterns in coyotes, beagles, and F_2 and F_3 generation coyote × beagle ("coydog") hybrids. The objective was to determine what effect, if any, domestication may have on prey killing in wild and domestic canids and their hybrids.

This investigation is divided into two parts, the first dealing with more qualitative aspects of behavioral organization and temporal ordering or sequencing and the second with quantitative analysis of two action patterns, the vertical leap and the forelimb stab. (For detailed descriptions of these action patterns, see Fox, 1969 and 1971b.)

Materials and Methods

PREY-CATCHING AND KILLING

In the qualitative part of this investigation, subects were 8–9 weeks old when tested and had no prior experience with live or dead prey. Differences in behavior might therefore be attributable to genetic rather than experiential influences. The temporal ordering of action patterns and the frequency of certain action patterns were recorded for qualitative and quantitative analysis, and tests were repeated in order to determine what effect experience might have on subsequent prey-catching and killing behavior. Additional studies were made on six coydogs in order to evaluate the possible influence of experience in the test/retest design of the experiment.

Four beagles (littermates), 9 F_2 (from 3 litters), and 8 F_3 (from 3 litters) generation beagle × coyote hybrids and 13 coyotes (from 3 litters) aged 8–9 weeks were used in this study. They were hand-raised in incubators from 3–6 days of age, being fed regularly on Esbilac via stomach tube, as described by Fox (1966). This method was chosen to control for individual differences in maternal behavior. All subjects were raised indoors in pairs or trios and had no exposure to live or dead prey prior to testing. They were never fed fresh meat and were raised on commercial dog food (Purina Chow and Gaines Burgers) being weaned at 3 weeks of age onto this food blended with Esbilac and housed in littermate groups in 4 × 10 ft cages. Each subject was carried from the animal room into the soundproof testing room opposite and was then observed alone in a 5 × 5 ft arena, where the prey, a 4 to 5-week-old white rat, was placed. This type of prey was chosen for convenience, being readily available and having been used consistently in earlier studies with canids. An arbitrary cutoff time of 20 min was assigned for each test, during which all behavior was recorded against a time-base using world-symbols and abbreviations for the various action patterns (see Tables I–III). It was then possible to quantify the frequency of certain action patterns and to trace temporal sequences of prey-catching, killing, dissection, ingestion, and play with prey. If the subject did not kill and/or eat the prey during this 20 min period, the prey was killed immediately after the test and was presented to the subject with the abdomen cut open to expose the liver and viscera, and observations continued for a maximum period of an additional 30 min. If the subject gave no response after 5 min, the liver was removed and placed in its mouth in order to sensitize the subject to the carcass. This is occasionally effective in stimulating a naive canid to eat prey for the first time (Fox, 1969). The action patterns associated with dissection and ingestion of the prey were recorded until the prey was eaten or partially eaten and ignored. This time period varied up to approximately 30 min.

All subjects were retested 24 hr later in order to determine what effects the first experience with prey had on subsequent organization and sequencing of action patterns.

VERTICAL LEAP AND FORELIMB STAB

From a qualitative study of prey-catching and killing in coyotes, beagles, and F_2 generation coyote × beagle hybrids, it was observed that two very distinct action patterns were a consistent part of all three of the above species' behavior repertoire. These action patterns are the forelimb stab and the forelimb stab with a vertical leap (see Figure 1). The latter is analogous to pinning down described by Eisenberg and Leyhausen (1972) in other predators.

It was felt that quantification of the frequency of the occurrence of these action patterns in controlled observational situations might render an appropriate index of the degree to which these three species were aroused and motivated to elicit these action patterns. The second section of this chapter, therefore, deals specifically with quantified analyses of these particular behaviors.

After completing the prey-killing tests at 9 weeks of age, the same litter of four beagles, two of the litters of six F_2 coydog hybrids, and one of the litters of three coyotes, were kept for this study. Four beagles (littermates), six F_2 generation beagle × coyote hybrids (littermates), and three coyotes (also littermates), aged 9–12 weeks of age were thus used in this study. Each subject was observed in the 5 × 5 ft arena, the floor of which was covered by a

Figure 1. *Vertical leap (a) followed by forelimb stab (b) directed at visually concealed prey by F_2 generation coyote × dog.*

white linen sheet. The edges of the sheet were secured under the plywood walls of the arena. The prey, two 4 to 5-week-old white rats, were placed in the arena under the sheet. The two action patterns (vertical leap and forelimb stab) were most often elicited when the canids observed the movement of the rats under the sheet. Each test was arbitrarily divided into two 10-min intervals (20 min for each test), during which the frequencies of these two action patterns were recorded. The total interaction time between canid and prey was also recorded. In addition, qualitative notes describing major events such as death of the prey and overt signs of degree of arousal of the subjects were also recorded. Each 20-min test was repeated within a week after the initial test.

Results

PREY-CATCHING AND KILLING

Beagles

In the four beagles tested, all subjects reacted toward the rat; but interactions were brief, and the rat would frequently be ignored as the dog suddenly began to investigate the arena or attempted to get out of the arena. These observations clearly imply that the beagles were less motivated than the coyotes and coydogs. No prey was killed in either the first or second tests.

The following behaviors were recorded: orientation, approach, follow, investigation (sniffing and pawing), approach/withdrawal, bite intention, nose-stab, head-shake intention, leap-leap followed by bite intention, nose-stab or head-shake intention or pawing. These actions may be followed by an immediate leap backwards or sideways. Only one beagle contacted the live rat with its teeth, the prey being licked and gently nibbled with the incisors. All subjects showed play-soliciting behaviors, including the bow, exaggerated approach/withdrawal, tail-wagging, and also barking and yelping at the rat.

In the first test with live prey, none of the following action patterns were observed: bite, carry, bite and head-shake, stab with forelimbs, hold with one or both paws and bite. No dog killed the rat in the first or second tests.

These observations show that although the dogs were attracted toward the rat, the duration of interaction (and therefore possibly the motivation) was much shorter than in coyote or coydog and a number of prey-catching and killing action patterns were absent. The normal canid temporal sequence of orientation, approach, investigation, bite, carry, bite (and head-shake) to kill, followed by dissection and ingestion was truncated at the bite interphase. Investigation was instead followed by self-play (chasing own tail), play-soliciting (which is not unusual in wild canids but which is normally satiated after 10–15 min), or by a bite intention. The dogs would lunge at the rat and bite *in vacuo* (air-snap) or deliver nose-stabs which were interpreted as bite intentions. Such actions were often preceded by forward leaps and followed by one or more backward leaps. This leap-leap-nose-stab sequence (see Table I) was only observed in the beagles and also occurred during intraspecific play. The same sequence of actions has been recorded in Chihuahuas, malemutes, malemute × wolf hybrids, and several domestic dogs of mixed breeding, but not in coyotes or wolves during intraspecific play. The possibility that this action pattern sequence is a specific marker for *Canis familiaris,* as a species characteristic, should be considered.*

When retested with live prey after 24 hr deprivation of food, only one of the four subjects showed no additional inclusions of the actions associated with the prey-catching and killing sequence. In one subject, the rat was bitten 13 times, and on each occasion the bite was inhibited, the rat being uninjured and active at the end of the test. The rat was grabbed and carried five times by this subject, and on the third carrying episode (15 min after onset of test) a weak head-shake occurred which was more violent on the fourth carrying episode but did not occur subsequently. This subject also made four low amplitude forelimb stabs at the rat during this second test. In the first test, this subject directed neither bites nor forelimb stabs at the rat. Another subject added only two inhibited bites to its earlier repertoire.

*More important though is the fact that this is an intraspecific action that was directed at the prey which, as emphasized earlier (Fox, 1969), may be responded to as a play-object or partner in some canids until it is learned that the prey is food.

Table I.
EXAMPLE OF INDIVIDUAL NOTATIONS

Beagle Male II (First Test)

No immediate orientation.

2 *min*	Then orients, rushes, paws, TW (tail-wag), sniffs, then ignores, then approaches, TW, bite intention, head-shake intention. Self-play leap-leap-leap-nose-stab.
5 *min*	Investigates then ignores rat.
6 *min*	Leap-leap-nose-stab, paws, chases, bite intention, investigates, TW.
7 *min*	Licks rat's face. Paws, bite intention, play-solicits, TW.
8 *min*	Lunge, nose-stab, A/W (approach/withdrawal), play-solicits. Nose-stab, A/W, bite intention.
10 *min*	Play-solicit, bite intention, yelp-growl, TW. (Brief poorly sustained interactions—no clear bite, carry or forelimb stab.)
11 *min*	Self-play.
12 *min*	Rushes, gently nibbles, paws, A/W, nose-stab, bite intention. TW, play-solicit, paws, yelps, leap-leap-A/W nose-stab, bite intention, yelp.
14 *min*	A/W, bite intention, TW. Leap-leap-leap-nose-stab, A/W, play-solicits. Leap-leap-leap away, bite intention, leap away.
15 *min*	Growl-howl-bark at rat. Play-soliciting dance—sustained playful interaction for last 5 min. (Rat is alive and well.)

Beagle Female II (Retest)

Immediate rush, paws, and *bites,* but bite inhibited. Then ignores rat.

2 *min*	TW, bite—carry, then dropped in one corner. Inhibited bite, then bite intention. Yowl-barks at rat + nose-stab.

Table I. (*continued*)

5 min	Approaches and mouths and leaps away.
	Leap-leap and mouths.
7 min	Mouths, leaps back, then bite intention.
	Plays with tail of rat.
	Self-play.
	Play-solicits + nose stab.
11 min	Leap bite (inhibited).
	Paws and mouths.
15 min	Ignores rat, little interest.
16 min	Carries, drops, one forelimb stab.
	Carries, weak head-shake.
	Carries, strong head-shake.
	Drops, mouths, carries, mouths, drops, forelimb stab.
18 min	Stalks and rushes, forelimb stab, growl.
	Bite intention.
	(Rat is alive and well.)

A third subject showed a marked increment of actions, giving 14 inhibited bites and 8 forelimb stabs and grabbing and carrying the rat 10 times. On the fourth carrying episode, piloerection occurred, and on the sixth episode (10 min after start of test) the first of three head-shakes, which subsequently increased in intensity, were noted.

An interesting and bizarre coupling of action patterns was recorded after the second grab-carry episode, which occurred 8 times. This was the inclusion of a high amplitude leap, which on the last three occurrences, was coupled with a head-shake but which occurred after the dog had leaped at, grabbed, and carried the rat. Instead of the usual temporal sequence of leap, grab and bite, head-shake and carry (or carry and head-shake), this subject coupled the carry component with an additional leap while it had the rat in its mouth. And on three later occasions when this misplaced leap was seen, a violent head-shake was recorded simultaneously.

It is logical to propose that this prey-catching leap (while the prey was in the dog's mouth) was a misplaced component of a disordered temporal sequence and that it subsequently entrained the head-shake component which normally occurs after a leap when the prey is secured in the dog's mouth.

When given dead prey with the viscera exposed after the first test, additional evidence of the dogs' inability to efficiently dissect the prey was obtained. Only one of the four subjects ignored the carcass completely. A second subject ignored the cut-open rat until the rat's liver had been placed in its mouth. This was ingested and the dog then immediately went over to the carcass and licked out the viscera. Excessive licking continued for over 4 min. Ths was followed by the dog repeatedly lifting up the carcass with the incisors; the forelimbs were not used to secure the carcass beneath the feet to facilitate dissection. This lack of coordination of forelimbs and teeth continued for approximately 6 min, when the dog lay down with the rat alternately between and under the forepaws as it indiscriminately chewed and mouthed the abdomen and appendages of the carcass. After 15 min, one forepaw was severed. Each time the pup pulled on the carcass, the latter would slip from under a forepaw. After 25 min, the head was removed and ingested.

A third subject immediately grabbed and shook the carcass, gave several crush-bites, and attempted to bury it in one corner of the arena. This subject ate the rat's liver when given it but subsequently ignored the carcass. The fourth subject also ate the liver but ignored the carcass.

After the second prey-killing test, all four subjects were more reactive toward the carcass. The first subject licked out the viscera and inappropriately used the carnassial teeth to separate the viscera. This took over 15 min. The normal reaction is to tear the viscera with the incisor teeth once the prey has been secured under the forefeet for counter-traction. This subject never secured the rat under the forefeet and subsequently ignored the rat after ingesting the liver; it appeared incapable of dissecting the carcass.

The second subject violently shook the dead rat as before and then appropriately crushed and severed the head with the carnassials and ingested it. The rest of the carcass was masticated with the carnassials, the forelimbs now being better integrated with teeth actions, in that the subject would frequently hold the carcass under one paw while chewing the rat with the carnassial teeth.

Instead of using the incisor teeth and counter-traction with the forepaws to tear the skin, limbs, and viscera, the third subject continued to masticate the carcass with the carnassials. Consequently, the carcass was not dissected or eaten.

The fourth subject first used the carnassials like the preceding subject, but after 6 min switched and used the incisor teeth in combination with forelimb counter-traction to tear and dissect the rat. The viscera were ingested followed by a long period of chewing the rat's appendages with the carnassial teeth. This latter ineffectual activity after 8 min was directed to the head of the carcass, which was crushed, severed, and ingested. The remainder of the carcass was subsequently ingested. These observations are presented in detail in order to demonstrate the role of experience in improving prey dissection and ingestion.

Coyotes

The detailed ontogeny of prey-killing behavior in the coyote has been described earlier (Fox, 1969), and only the major points relevant to the present investigation will be presented (see Table II).

In one litter of four 8-week-old coyotes, all immediately grabbed and bit the rat, the bite being uninhibited. The coyotes immediately killed the rat and had dissected and eaten it within 1 and 3 min, respectively. The other two killed their prey after 2.5 and 5 min, respectively, each taking a further 2.5 and 7.5 min to dissect and eat the carcass. The slowest of these four subjects engaged in excessive bouts of head-shaking which was released every time the prey moved in the coyote's mouth. Long bouts of violent head-shaking continued during dissection of prey; when later retested, this same coyote engaged again in bouts of excessive head-shaking, but the bouts were less frequent than on the first test. All subjects when retested showed improved orientation of the grab-bite, striking the prey in the anterior thoracic region rather than aiming more posteriorly, which would give the rat sufficient freedom of movement to bite the subject's face. When retested all four subjects had killed and ingested the prey within 60 sec (aver-

Table II.

	Coyote Female II, Group II (First Test)
1.5 min	Immediate grab-bite-carry, then crush bite. Was bitten on face—released rat, then grabbed around neck and thorax + violent head shakes. Multiple crush bites.
2.5 min	Crushes TW, A/W, TW, crush bites, A/W, TW. Crush bites, carries. A/W, TW, carries—intermittent play and crushing.
4.5 min	Pulverizing, holds with forepaw. Crushes head with carnassials, holding body with one forepaw.
7.5 min	Eats off hind leg, then tail.
11 min	Eats off other hind leg. Uses carnassials and forelegs well. Rips skin with incisors using forelimb counter-traction.
13 min	Eats intestines.
14 min	Uses carnassials and removes and eats head.

age 35 sec). None of these four coyotes played with the live or dead prey. All showed perfect coordination of forelimbs and teeth in pulling the carcass to pieces and used their teeth appropriately— the carnassials to shear and crush and the incisors to pull and tear, in contrast to the beagles described earlier (that lacked forelimb coordination and used their teeth inappropriately, such as using the carnassials to tear the skin and viscera).

In a second litter of six coyotes, the average killing time was 3.3 min, all subjects immediately orienting and grabbing the rat with no bite inhibition. Usually multiple crush bites and interspersed head-shakes killed the rat. Two subjects showed very brief play-soliciting and tail-wagging, and at such times the bite was inhibited. All subjects threw the rat across the arena with a powerful upwards and sideways head flick, the frequency of this being greater in the two subjects that engaged in brief bouts of play. All showed high amplitude forelimb stabs which were often preceded by a vertical leap. Only one subject did not ingest the dead prey, even on retest.

When retested, all subjects spent less time chewing and severing the appendages of the prey and assumed the normal sequence of crushing, severing, and ingesting the head, then crushing and tearing the rest of the carcass, pieces of which were ingested at random, usually the pelvic region and hind legs being ingested last; the caecum was rejected by two of the subjects.*

F_2 Coydogs

All coyote × dog hybrids showed varying periods of play with prey, and all had clearly inhibited bites. Leaping and catching patterns were well coordinated and integrated, but the killing bite was absent. Most subjects toward the end of the first test, as in the beagles, developed more intense head-shaking, but in contrast to the beagles, the grab-bite-carry sequence was present at the beginning of the test. As with the coyotes, these hybrids never vocalized, while all beagles barked, yelped and growled at the live prey.

The doglike leap and nose-stab was also seen in all six F_2 hybrids. They would throw the live prey across the arena with a head flick and then leap upon it with a coyotelike high amplitude leap and forelimb stab. Only one of these subjects killed the rat after approximately 10 min, and, in fact, no killing-crushing bite was applied; the rat was essentially eaten alive, and on retest, this same occurrence was recorded. This clearly underlined the omission or inhibition in the coydog hybrids of the killing bite, although all other prey-catching patterns were evident (see Table III).

As the tendency to shake the live prey increased toward the end of each test and was much more frequent in the second test, so the inhibited bite and multiple nibble-bites became more intense. In the second test, only one other coydog killed the prey after 8 min, and the multiple killing bites (combined with violent headshakes) appeared to be disinhibited nibble-bites. This subject subsequently ignored the dead prey. The frequency and duration of play bouts were less in all subjects in the second test, yet unlike the

---

*These prey-killing tests were arbitrarily conducted when the subjects were 8 weeks of age. Interestingly, a third group of three coyotes tested at 52 days of age did not kill on the first test. All showed the full prey-catching sequence (including violent head-shakes in two subjects), but in each case the killing bite was inhibited. All killed prey, however, when retested later. Further studies are therefore indicated to determine the maturation of prey-catching and killing prior to 8 weeks of age.

Table III.

Coydog, "Pale" Male (Retest)

Immediate chase, grab-bite, carry, then investigates. Chases, carries, holds in paws, TW, bite intention, carries, mouths.

5 *min*   TW, nibbles (no blood). Play-soliciting posture, paws, nibbles, carries, weak head-shake.

6 *min*   Bite clearly inhibited; play-soliciting and TW combined with grab-bite, head flick and pawing and carrying. Bite becoming less inhibited.

8 *min*   Blood, mouthing, holds with forepaws, paws, frequent nibbles, paws, rat bites pup on lip—distress vocalization. Then more violent play, head flick, leap, grab and carry.

9 *min*   Holds with forepaws and begins to eat leg. Rat still alive. Now A/W, TW, play leaps, briefly, then chews and pulls forelimb, holding rat down with forelimbs. Rat dying.

13.5 *min*   Now eating rat proper—eats head first.

---

beagles, interaction with the prey was continuous (i.e., as in the coyote, motivation or arousal was more intense than in the beagles). One subject in the first, and this same subject and two others in the second test ingested dead prey (presented with the abdomen cut open), all three animals showing perfect coordination of head pulling and forelimb counter-traction to tear the prey. The incisor and carnassial teeth were used appropriately to tear, crush, and shear the carcass, and all subjects followed what is thought to be the normal canid pattern of crushing, severing, and ingesting the head first.

The above observations are summarized in Table IV which serves to illustrate the general trends of the three groups of canids studied in this investigation.

In order to control for the interaction effects between test and retest of prey-killing, where after the first test the subject was presented with dead dissected prey, a second group of three 6-week-old F_2 coyote × dog hybrids were tested as follows: three

Table IV.

NORMAL TEMPORAL SEQUENCE OF PREY-CATCHING AND KILLING
IN COYOTE
(With point of truncation of sequence in beagles of F_2 coyote × dog hybrids)

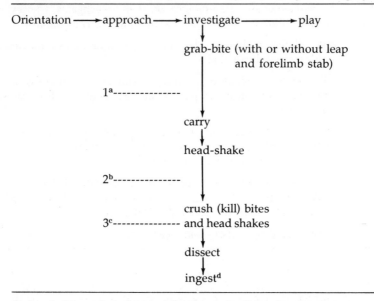

[a]1, Level of truncation of temporal sequence in all four beagles, first test.
[b]2, Level of truncation of temporal sequence in all four beagles, second test.
[c]3, Level of truncation of temporal sequence in all F_2 coydogs, first test, the kill-bite being inhibited in four out of six subjects.
[d]Complete follow-through of temporal sequence occurred in only one coydog, and in this subject there was no clear kill-bite.

consecutive days of testing, each test lasting 10 min, and dead dissected prey being given not after the first trial as in earlier studies, but after the second trial day. In this way, the contribution of experience with live prey in Trial I influencing performances on Trial II could be separated from the possible added effects of experience with dead prey between Trials I and II.

The data are summarized in Table V. An overall decrease in response latency over successive trials is evident. In S_1, the bite

Table V.

CHANGES WITH SUCCESSIVE DAILY TRIALS
IN THREE F_2 COYDOG HYBRIDS

Trials	Actions patterns	*Response latencies of subjects (sec)*		
		S_1[a]	S_2[b]	S_3
I	grab-carry	245	45	325
	head-shake	—	—	325
	crush-bite	—	—	—
II	grab-carry	30	10	5
	head-shake	—	—	5
	crush-bite	—	—	—
III	grab-carry	10	45	15
	head-shake	—	360	15
	crush-bite	—	380	—

[a]Ate dead, dissected prey at end of test.
[b]Partially killed and ate live prey.

was inhibited in all trials and no head-shakes were recorded, although the actions of grab-bite and carry, stalking, leaping, and play-soliciting were recorded. Occasional head-shake intentions while approaching the rat were recorded in this subject in Trials II and III. The subject handled dead dissected prey in an uncoordinated manner resembling that of the beagle, and the general phenotypic features of this F_2 hybrid were closer to beagle than coyote. S_2, which phenotypically resembled more the coyote, killed its prey on the third trial after 8 min; the head-shake first appeared in this trial after a long latency, and as observed in the other hybrid group, there was a gradual disinhibition of the bite, multiple nibble-bites increasing in intensity until the prey was crushed around the thorax. This subject, as also observed in the first group of hybrids, began eating the prey while it was still alive. Both S_2 and S_3 showed good coordination of actions in dissecting and ingesting dead prey. Throughout all trials S_3 showed a strong head-

shake action, but the bite was inhibited and all prey were uninjured at the end of each trial. Clearly, the experience with dead
prey between Trials II and III had no appreciable effect on the
prey-killing tendencies of S_1 and S_3. Such experience may have
contributed to the development of crush-bites which almost killed
the prey of S_3, but another interpretation can be offered, namely,
the increasing arousal, short response latencies and increasing duration of interaction over successive trials and the close temporal
linking between head-shaking and crush-bites in this subject. Further studies are needed in order to determine the relative importance of maturation, increments of experience with live prey, and
the effects of ingesting dead prey on the development of preda-
torial action sequences.

F_3 Coydogs

None of the eight F_3 generation coyote × dog hybrids from
three litters killed prey on the first test, and when retested, only
one killed and ingested its prey. With this one exception, all subjects showed only brief interest in prey during the first test, reactions being orientation and following but no contact. When retested, additional reactions appeared, grab and carry in two subjects and head-shake and bite intentions in another. Reactions in
the third trial were virtually the same. The one F_3 hybrid that killed
after a latency of 6.5 min on the second test was the most responsive in the first test. It immediately grabbed and then carried the
live prey and after 7.5 min gave a mild head-shake. After this it
continued to play intermittently with the prey until the end of the
test period.

Clearly, the F_3 hybrids have a significant repression of prey-
catching and killing behavior. This could be because they were
more disturbed than the other canids when in a relatively unfamiliar place (i.e., the testing arena). In order to assess this possibility,
they were tested again in their home cages. (The day prior to this,
each litter of three pups was placed together in the arena with live
prey. It was virtually ignored, subjects engaging in social play instead.) In each litter a tug of war began over the dead prey placed
in their cage. It was not eaten, however, and was treated like a play
object. With live prey, one litter did not kill in its home pen, while

one individual in the second litter killed the rat after 10 min; multiple nibble-bites to the body became disinhibited and suddenly the thorax of the prey was crushed. The rat was then eaten. Although the prior group test in the arena with prey did not reveal any social facilitation of prey-killing, social facilitation (i.e., competition) may have triggered prey-killing in the home pen, as emphasized by Leyhausen (1973) in his studies of prey-killing in cats. The significance of a particular place to eat and hunt is worth further investigation. Burrows (1968), for example, records that red foxes may not kill prey within their denning area, and a rabbit or ground-nesting duck may even share the same denning site. The same animals would undoubtedly be killed if found in the fox's range when it was out hunting.

RESULTS OF FORELIMB STAB AND LEAP-STAB OBSERVATIONS

In the four beagles tested, all subjects interacted with the prey moving under the sheet. However, the interactions were of very brief duration, and all four subjects never exhibited either vertical leaps or forelimb stabs. The quantitative results of all tests are reported in Table VI. In most trials the beagles spent a great deal of time in both play-soliciting and self-play behavior. Frequently a subject would completely ignore the prey for most of the test and then finally spend a minute or two growling, barking, and nibbling at the prey. All subjects would claw with their forepaws at the prey when the prey was lying immobile against the side of the observation arena. When the prey attempted to escape from the beagle, the subject's most intense response would be a nose-stab, whereas the other two species would respond with a forelimb stab or a vertical leap and stab. There was no significant difference found to exist between Trials 1 and 2 for the beagles' scores for forelimb stabs and forelimb stabs with vertical leap.

In contrast to the beagles, the three coyotes tested showed high frequencies of both action patterns and on the whole the coyotes spent significantly more time interacting with the prey. The coyotes followed every movement of the prey; they would stalk

Table VI.
DATA FOR TESTS OF BEAGLES, COYOTES, AND F_2 COYOTE × BEAGLE HYBRIDS[a]

Actions	Mean & Range	Coyotes (n=3)		Hybrids (n=6)		Beagles (n=4)	
		Trial 1	Trial 2	Trial 1	Trial 2	Trial 1	Trial 2
Forelimb stabs	Mean	8.67	10.33	17.76	13.67	0	0
	Range	4–12	7–13	9–29	7–28	0	0
Forelimb stabs + vertical leap	Mean	5.33	9	11.67	8.33	0	0
	Range	2–7	7–11	3–25	2–12	0	0
Interaction time	Mean	10:31	10:14	12:26	8:37	4:11	6:23
	Range	9:37–12:12	8:30–11:31	9:47–15:22	5:17–11:24	1:11–7:10	1:30–10:16

[a]Frequencies of action patterns in 20-min trial and duration of interaction with prey in minutes.

and chase the rats, trap them, and make uninhibited bites, head-shakes, and several forelimb stabs and leaps. One coyote actually killed the prey. In the two tests, the prey was killed with a crush-bite, and then the coyote attempted to pull the dead rats through the sheet. In three other subsequent retests, the prey were apparently the victims of the forelimb stab with vertical leap. In these cases, the subject ignored the rats after they were killed.

Among the F_2 generation beagle × coyote hybrids, there was a high frequency of both forelimb stabs and vertical leaps, and the total interaction time between subjects and prey was higher than in both the beagles and coyotes. (See Table VI.) All of the hybrids required approximately 2–5 min to adapt to the observation arena. In contrast, the beagles sometimes spent the first 10 min adapting. This adaptation period consisted of the following types of behavior: initial investigation of the arena (ignoring the prey), several attempts to escape from the arena, self-play (tail-chasing and jumping in the air), and finally encounter with the prey, usually by accident. Once the F_2 hybrids did encounter the prey they became highly aroused. Most subjects exhibited a high frequency of forelimb stabs. These were usually elicited by small movements of the prey. In contrast to the coyotes, the beagle × coyote hybrids never used a crush-bite on the prey (emphasizing this same point observed in the previous test with live, visible prey). All the hybrids' oral contact with the prey consisted of either inhibited biting or multiple inhibited nibble-bites. As in the case of the coyotes, the coydogs exhibited a high degree of arousal and motivation as evidenced by the high frequencies of forelimb stabs and leap-stabs, long bouts of intense stalking behavior, and long periods of interaction time with the prey. The hybrids also displayed a high incidence of what may be referred to as place memory. For example, the subject would be interacting with one of the two prey, then would move to the other rat some distance away and then, after some time, suddenly turn and make a high amplitude vertical leap directly at the spot where the first prey had been located. Many of the leap-stabs were observed in this context or when one of the hybrids was chasing a fast escaping prey. In the latter case, the subject would make one or two forelimb stabs at the prey; the prey would then start to flee, at which point the subject would make a vertical leap and stab.

Discussion

The two related studies bring out a number of intriguing points, the genetics of which can only be speculated upon at this stage in view of the small sample of coyotes, beagles, and hybrids used in this project. More precise genetic inferences may be drawn in future studies involving not only a larger n but also F_1 and back-cross coyote × dog hybrids. In spite of these limitations, qualitative and quantitative differences in prey-catching and prey-killing were remarkably consistent within groups and add to our understanding of the effects of domestication on canid behavior.

The most striking and consistent findings were the absence of bite inhibition in the coyotes in the two tests where the prey was either visible or visually concealed and complete inhibition of the prey-killing bite in the beagles. In contrast to these extremes, two of the F_2 hybrids showed a gradual release from inhibition, and although prey were eaten, they were not as effectively killed as by the coyotes.

The difference in the degree of bite inhibition may account for the truncation of temporal sequences of prey-catching and killing, but it should be remembered since temporal organization is varied during play, that coyotes will play with live prey for variable periods and at such times show clear bite inhibition prior to killing the prey (Fox, 1969). The truncation of the sequence may be attributable to differences in arousal in the beagles and hybrids compared to the coyotes, a high degree of arousal being necessary for actions such as crush-(or kill) bite, head-shake, forelimb stab and leap-stab to be elicited.

Another related hypothesis is that the action patterns associated with prey-catching and killing are ordered temporally on the basis of threshold differences, those actions toward the end of the sequence having a higher threshold than earlier actions. A minimally aroused subject would then only show low threshold responses such as orientation, approach, and follow, as was the predominant feature of the beagles in the first test. Arousal (or motivation) may therefore increase with experience with prey, so that higher threshold action patterns are incorporated appropriately into the temporal sequence. This hypothesis is supported by the retest data from beagles with visible prey.

The motivational state may therefore change with experience and may vary between individuals and between species and hybrids; differences in conflicting motivations, such as the tendency to play or to be fearful of the test situation as emphasized by Mason (1967), must also be considered.

It may be postulated, therefore, that optimal arousal and motivational conditions are essential for the release of certain actions before experience per se can have any effect upon them.

The inferior prey-killing ability in beagles and in some of the coyote × dog hybrids may be due to a delaying effect (of domestication) in the maturation of this behavior. Thus, it may be argued that the age of 8 weeks may be too early to test prey-killing in beagles and certain hybrids. Such animals when retested at 10 and 12 weeks of age did not, however, show any further improvement in prey-killing ability, and it is unlikely, therefore, that there is any delay in maturation per se, but rather, as implied from this study, there is a change in threshold and truncation of the temporal organization and sequencing of action patterns attributable to domestication.

Analysis of the two action patterns, forelimb stab and leap-stab, reveals that these actions are a predominant part of the behavioral repertoire of the coyote and coydog hybrid, but under the test conditions (concealed prey) they were not reported in the beagles. This apparent absence in the beagles may be correlated with their low overall duration of interaction scores, which indicates that they were less aroused than the coyotes and hybrids. The fact that in one beagle leap-stabs were occasionally recorded when retested with live visible prey supports the arousal hypothesis alluded to earlier in this discussion. A second possibility is that the forelimb stab and leap-stab occur at a lower frequency in the beagles because they have a higher threshold; a greater degree of arousal is needed for its release than in the coyotes and hybrids.

Bite inhibition was clearly manifest in the beagles and their coyote hybrids, and inhibition could not be attributed to motivational factors alone since the hybrids were highly motivated even though they showed bite inhibition. This notion is supported by comparative developmental studies of early contactual and aggressive behaviors in coyotes, wolves, dogs, and wolf × dog hybrids (M. W. Fox, personal observations) and prey-killing abilities

(Fox, 1969). Around 24–30 days of age, contact with a conspecific frequently releases an uninhibited bite in coyotes (and also in young red foxes of this age), and it is only after dominant–subordinate relationships have been established that the bite becomes inhibited. Prior to this time, play behavior is rarely seen. In contrast, wolves and dogs show marked intraspecific bite inhibition and also inhibition of prey-killing bite, which in part accounts for the fact that they engage in sustained bouts of contactual and play behaviors compared to coyotes and foxes (Fox, 1975). The conclusion, then, is that there is a greater degree of genetic control of bite intensity in domestic dogs (beagles), while in coyotes selective control is effected more through social and experiential influences.

In some breeds of domesticated dogs, notably in the bird dog pointers and retrievers, there has been rigorous artificial selection for "soft mouth," for the dog must not mark the game. The prey-catching sequence in such breeds consists of orientation, approach (or track), and either point or grab and carry (retrieve), the killing bite and head-shake actions being selectively eliminated.

Vauk (1953) reports that play with prey (and with inanimate objects) develops at different ages in different breeds of dog. He concludes that domestication has modified or completely inhibited prey-killing, in certain breeds; some, like the pointer, develop fixed, exaggerated signs. The effects of selective breeding and training on prey-catching and killing behavior in the domestic dog is summarized in Table VII. It is evident that domestication and selective breeding may lead to a breakdown in the normal temporal sequence at one of several points by intensifying or inhibiting certain action patterns. Various breeds, selected for specific hunting and tracking tasks, clearly demonstrate this subtle aspect of domestication. The capacity of domestic dogs to hunt and kill prey and live independently has been demonstrated in feral dog studies (see Chapter 3). Therefore, these abilities are not eliminated through domestication, but rather, motivation, threshold of different actions, socialization, training, and dependence upon man are some of the interrelated variables that must be considered in evaluating the effects of domestication upon animal (and human) behavior.

Two 8-week-old Australian dingos (undomesticated) were recently tested. All killed and ingested prey on the first test, with an

Table VII.

Complete hunting sequence of wild canid	Partial sequence of some domesticated dogs[a]
Tracking, trailing	Bloodhound, gazehound
Herding, driving	Sheep dog
Stalking, pointing	Setter, pointer
Attacking, killing	Boarhound
Retrieving (for cubs/mate)	Retriever

[a]Attack inhibited in most breeds.

average kill latency of 15.5 min. When retested, this latency dropped to 10.5 min. Interestingly, one pup, prior to killing, held the prey with its forepaws and pulled at its legs and tail, the consummatory (eating) phase occurring before the killing bite.

This clearly demonstrates that inheritance does influence bite inhibition—a point of considerable social significance in view of the number of people being bitten by dogs today (estimated 400,000 per year in the United States). While improper human handling may often be to blame and/or inadequate socialization (Fox, 1972), a relaxation of selection for stable temperament and bite inhibition may also be involved especially in those breeds that become popular and are mass-produced, with little quality control, to meet public demand. The release from inhibition in well-socialized dogs for attack training can be effected by increasing arousal through play-fighting with a padded sleeve. A gradual disinhibition occurs without disruption of the social bond. This unlearning of a socially inhibited response (biting) is analogous to the gradual appearance of the killing bite in coydogs where motivational factors or social inhibition per se may be involved. The reaction of beagles toward live prey was essentially one of social play and logically, social inhibition of biting occurred. This conclusion was also made in earlier studies of prey-killing behavior in wolf cubs. Those that did not kill live prey showed playful and actively submissive actions toward the prey, but after being forcibly fed dissected prey, they essentially learned that it was food and from then on would attack and kill prey and direct no further social behavior toward it.

Thus, some species (wolves) have to learn what to kill and what is food, while in others (foxes and coyotes) learning has less influence (Fox, 1969). The argument here is that there may be some connection between intraspecific aggression, general bite inhibition, and prey-killing. By analogy also there may be considerable resistance and conflict in training a dog to selectively attack certain people (on command or without command in a certain place they may guard) and not to attack other people or in other places, when there is socialized attachment to people and generalized bite inhibition. Therefore, no single factor is wholly responsible for prey-killing and social inhibition of biting in wolves or of selective disinhibition of attacking and biting man in trained guard dogs. Genes, early experience, social attachment, and training all contribute to the ultimate expression of these motivationally separate activities of prey-killing and agonistic behavior. Social attachment to prey (Kuo, 1960) may inhibit prey-killing behavior; intraspecific bite inhibition may be associated with a more generalized bite inhibition which has to be disinhibited in relation to prey in some species such as the wolf. Thus, although motivationally distinct, certain variables may affect both aggressive behavior and prey-killing reactions equally or in a similar direction; hence perhaps the frequent but incorrect association between hunting and aggressive behavior.

Similarly, wolf cubs show more social inhibition of aggression and biting conspecifics than coyotes and foxes who at an early age may bite a conspecific reflexively as though its fur and movement released a prey-killing response. Social experience in controlling bite inhibition in coyotes and foxes is therefore important in early life, while release from a more generalized inhibition of biting is necessary for some wolf cubs to secure live prey.

In future studies it would be advantageous to use pressure transducers to provide quantitative data on developmental changes and on species and breed differences in bite intensity and control. The contribution of genetic and experiential influences might then be elucidated. Ewer (1973) notes that the canine teeth are highly innervated in carnivores to provide tactile feedback so that the positioning and pressure of the bite can be exactly controlled.

Summary

In order to identify possible effects of domestication on behavior, prey-catching and killing behavior was studied in domestic dogs (beagles), coyotes, and F_2 and F_3 generation beagle × coyote hybrids. The full temporal sequence of prey-catching and killing, together with efficient dissection and ingestion of the prey evident in coyotes at 8 weeks of age, were respectively truncated and disorganized in the domestic dog and hybrids. These changes in behavioral organization and temporal sequencing of action patterns are discussed in relation to differences in arousal (motivation) and response threshold. Partial and complete inhibition of the killing bite was found in hybrids and beagles, respectively, this being a major factor in the truncation of the full prey-catching and killing sequence. Domestication may lead to truncation of this temporal sequence at one of many points by intensifying or inhibiting certain action patterns through selective breeding.

VII

Interspecies Interaction Differences in Play Actions in Canids

Introduction

Most studies of social behavior have focused upon intraspecies interactions. This study involves a little explored area of interspecies social interactions, where the frequency of occurrence of clearly identified discrete action patterns may be modified in relation to the degree of dyadic compatibility within the selected context of playful interaction.

Yves Rouget (personal communication) raised a red fox (*Vulpes vulpes*) with a domesticated dog. The impressive, but unquantified, consequence which he recorded on film was the high frequency of face-oriented pawing manifest by the fox. Such an action is rare in foxes but is a common action pattern in domesticated dogs. This kind of observation opens the question of the role of experience in the development and reinforcement of species-typical action patterns.

An incompatibility of reciprocal actions has been reported by Blauvelt (1964) between sheep and goats raised together; no effective fights were seen because the sheep tried to butt the goat and the goat tried to jump on the sheep. Interestingly, whom the animal attacked was determined by early social experience, but the manner in which it attacked remained unchanged.

A consistent feature of social play in *C. familiaris*, F_1 *C. familiaris* × *C. lupus*, and F_2 *C. familiaris* × *C. latrans* is the occurrence of the play-leap (Figure 1). This action may be preceded by a direct stare, a play-soliciting bow or an incomplete leap intention which may be repeated three or four times. The leap itself varies in amplitude and may be repeated as the subject moves in on its congener. The latter may also rear up and meet its partner with a reciprocal leap, and both animals briefly stand and wrestle with the forelimbs and make face-, cheek- and scruff-oriented bites or incomplete bites. Most often the leap is followed by an incomplete bite or bite intention or by a nose-stab which may be interpreted as a bite intention movement. Occasionally this may be followed by a backward leap.

In the coyote, *C. latrans*, such sequences involving the play-leap are not observed. In this species, a play-soliciting bow is fol-

Figure 1. (A) *Vertical leap of beagle during play which may be reciprocated (B) and both rear up. Leap may be followed by face- or cheek-oriented bite. (D) Playful standing over by F_2 coyote × beagle; (E) clasping by coyote during play-fighting.*

lowed by side-to-side head flexions which move fluidly down the body, followed by rolling over, "spinning," running around the partner, "diving" or exaggerated approach or withdrawal (Fox, 1971b). From the play-soliciting bow, the head may be twisted upward toward the forelegs, throat, or cheeks of the congener and bite intention lunges or incomplete bites are then executed (Figure 1).

An action resembling the play-leap occurs when one coyote is chasing another; the action is of low amplitude and is followed by a bite or bite intention on the rump or shoulders of the partner or by a forelimb clasp (Figure 1). This clasp, which is often never fully executed, consists of seizing the partner around the waist and frequently following with a scruff-oriented bite. The partner may

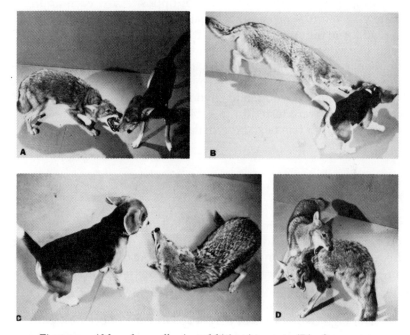

Figure 2. *Although scruff-oriented biting in coyote (D) often occurs during play-fighting, the coyote usually approaches low (A and B) and makes an upward directed bite from the (C) play-soliciting posture and rarely directs vertical leaps at its partner.*

attempt to avoid being clasped by executing a hip-slam or twisting around to face its congener. When facing, rearing up may occur, followed by face-oriented bites or bite intentions and pushing with the forelimbs.

Materials and Methods

With these species-characteristic actions in mind, it was decided to evaluate the compatibility and reciprocity of such action sequences in variously combined pairs of subjects, by recording the frequencies of play-leaps. All dyads were of opposite sex, each approximately 12 months of age. Each pair (of noncage mates) was observed for 10 min play in an 8 × 8 ft soundproofed arena equipped with one-way observation windows. Two observers recorded the frequency of play-leaps for each pair, dyad combinations consisting of beagle × beagle, beagle × coyote, beagle × coydog, coyote × coyote, coyote × coydog, and coydog × coydog. A total of 20 dyad combinations were tested in 12 subjects (4 beagles, 4 coyotes, 4 F_2 coydogs, with 2 females and 2 males per species).

Results

The findings are summarized in Table I. It should be noted that both coyote and coydog showed hip-slams during play, but the frequency of this action was not recorded. Also, several of the low amplitude play-leaps recorded in the coyote were difficult to distinguish from the incomplete action of rearing up to clasp the partner with the forelimbs.

The high frequencies of play-leaps in pairs of beagles, pairs of coydogs, and beagle × coydog reflect the reciprocal nature of this action. In all dyads where one individual scored over 15 play-leaps, then its partner likewise had a comparably high score.

The beagles were more active when paired with a coyote or

Table I.
SUMMARY OF PLAY-LEAP SCORES

Subject	With beagle	With coyote	With coydog
Beagle	8.5[a] (2–16)[b]	16.3 (1–37)	15.7 (2–35)
Coyote	1.0 (0–2)	2.5 (0–7)	0 (0)
Coydog	12.7 (1–31)	5.7 (4–9)	16.0 (7–22)

[a]Average scores per individual.
[b]Range shown in parentheses.

coydog than with its own species and this may account for the lower scores in beagle pairs.

An intriguing finding was that the coydogs had high frequencies of play-leaps when paired with a conspecific or beagle but lower frequencies when with the coyote. This was attributed not to a lowered activity or motivation, for play was intense, but to the possibility that the hybrids had a more flexible or adaptive play repertoire which could be reciprocally matched with a beagle or coyote partner. In contrast, the beagle when paired with a coyote showed the highest average frequencies of play-leaps which was not reciprocated. No significant differences attributable to sex were evident in any of the three canid types.

Discussion

These data support the general impression that the action sequences of beagle and coyote during social play are incompatible. The bow followed by upward directed bites and bite intentions of the coyote may be regarded as low ventral play-attack; the play-leaps of the beagle may be regarded as a high dorsal play-attack. The coyote, in responding to the ventral low attack orientation, twists away and usually executes a shoulder or hip-slam to block or deflect the attack; such actions were not observed in the beagle but were evident in the coyote × dogs and most probably contributed to their ability to maintain reciprocal synchrony with the coyotes during play bouts. These hybrids also showed the play-leap action which enabled them to sustain reciprocal play actions with the beagles.

The role of early experience and of social factors should also be considered, and this point is emphasized here rather than earlier in this book, with reference to the prior experiences of our canids. The criticism that might be leveled at this study to the effect that the beagles had no prior opportunity to interact and modify their actions with respect to coyotes (and vice versa) is untenable. All beagles and coyotes used in this study had been reared together from approximately 8 weeks of age, although tests were conducted using individuals from different cages. It might be expected that coyotes with such early social experience might acquire the play-leap action of the beagle. But this was not evident; thus supporting the notion that the play actions studied are species-characteristic action patterns, the frequency and temporal sequencing of which may be modified contextually as in certain dyadic interactions.

Given that the interaction with a different species may modify the occurrence and frequency of certain action patterns, more profound and long-lasting effects on interspecies interactions under conditions of sustained/enforced contact in captivity remain to be evaluated. While there may be little or no change in the structure (movement patterning) of certain action patterns, their occurrence and frequency of elicitation may be altered, especially within the broader context of domestication, as when different species (cat and dog; goat, sheep, and cattle; cattle and horses, etc.) are raised and confined together. Similarly, the influence of human behavior on such actions remains to be evaluated. One classic example is the canid analogue of the human "greeting grin" (which is quite distinct from the submissive grin and open mouth play-face of canids (Fox, 1971b). This facial expression is mimicked by the dog and has only been observed in man–dog interactions and not between dog and dog. There is also evidence that the capacity to elaborate this facial display is inherited since certain lines of dogs will never develop this behavior in spite of training, while others from the same kennel whose parents have this facial expression will quickly develop it either spontaneously or with a little human reinforcement.

Summary

Species-typical patterns were identified in wild and domesticated canids during play in dyads of the same or different species. While these actions are under relatively rigid genetic control (i.e., are inherited) other dimensions of the play sequence may be more easily modified contextually. Thus, not only species and context determine the frequency of occurrence of certain actions but also the compatibility of the interactee, be it of the same or of a different species. The long-term consequences of interspecies interactions under conditions of captivity may lead to more permanent changes in the occurrence, frequency, and amplitude of certain species-typical, "fixed" action patterns. This may be an additional variable to consider in investigating the complex influences of domestication upon behavior and warrants further investigation.

VIII
Socialization Patterns in Hand-Reared Wild And Domesticated Canids

Introduction

This chapter offers a number of observations about experiences with hand-raised and socialized canids, including wolves, coyotes, golden jackals, and red foxes. When assembled, these anecdotes constitute a long-term experiment on these species in captivity, where their social reactions toward people can be contrasted with their natural or innate social tendencies. The species studied belong to three basic social types (Fox, 1975); namely, solitary Type I (red fox), transitional Type II with permanent pair-bond (coyote and golden jackal), and the social Type III with permanent allegiances and pack formation (wolf). Temperament, a constellation of primarily inherited tendencies, interacting with experiential factors during early life, is a major determinant of social behavior and of the capacity to form and maintain social relationships in later life. In this chapter, it will be shown how the innate determinants of social relationships in these three canid types influence their reactions toward people after they have been hand-raised and socialized to human beings.

Materials and Methods

The following observations are based on experiences with 15 wild canids, all hand-raised from 2–6 days of age and kept in captivity under similar conditions of handling and management until sexual maturity (2 years) and in some animals up to 4 years of age. Subjects include 1 male and 2 female wolves, 2 male jackals, 2 female and 1 male red fox, and 5 female and 2 male coyotes. In addition, 6 beagles, 8 pointers, and 4 Chihuahuas were raised under similar conditions for comparative purposes.

THEORETICAL PREMISE

Observations are based on the reactions of these animals, all with known life histories, toward the author who hand-reared them, and toward research assistants who later took care of them. Over a 4-year period, the subjects had almost daily contact with the person who had raised them and also regular contacts with an average of one male and two female assistants per week, who were responsible for cleaning and feeding them. The turnover rate of personnel was approximately two every 12 months. The social conditions in terms of frequency and variation of human contact were therefore relatively low and constant. It should be added that those students studying the animals and this form of nontraumatic contact involved an average of two different observers per year. Occasional visitors were of course allowed into the facility on a random basis. In summary, all animals had regular contact/exposure to approximately six new people (sex ratio 1:1) each year. It is impossible to state whether this pattern of human contact facilitates secondary socialization or whether the annual turnover eventually leads to social confusion in some of the animals. Prior experience with such patterns with domestic dogs showed that most dogs readily accept new personnel. In contrast, dogs and coyotes raised exclusively by one or two people will often not accept a change in personnel or change of sex of handler and will show fear or aggression toward the new person (personal observations). In the present chapter the patterns of human contact in no way resembled the aforementioned restricted rearing and socializing conditions.

The distinctions made by Scott (1968b) between primary and secondary socialization should be emphasized here. Primary socialization refers to those social relationships which are established early in life during a critical period (Scott, 1962) as between a wolf cub, its parents, and littermates. Secondary socialization refers to those social relationships that develop at a later age, as between other adults of the pack that breaks up during the denning season. The possibility that species differ in their capacity to establish and maintain secondary social relationships will be explored in this chapter.

Primary social bonds may facilitate secondary socialization; a domesticated dog, for example, if raised with human beings early in life is able to generalize from these primary socializing experiences and is friendly toward strangers at a later age. Some breeds, notably the guard dog types, seem to have a more limited capacity to form secondary social relationships and are very wary of strangers. The relationship between this phenomenon and the comparable behavior in the wolf will be discussed subsequently.

Woolpy and Ginsburg (1967), in their extensive studies of wolf socialization, have shown that adult wolves caught wild, with careful handling, will become socialized to their handler and on the basis of such experience tend to generalize and respond socially to strange people as well. These findings put a different light on the critical period hypothesis but in no way refute it. Ginsburg reports that coyotes do not generalize or form secondary social relationships to the same extent as wolves. Instead they tend to remain attached to the principal person responsible for effecting primary socialization. This is the key to the present report; namely, to what extent do the socialization capacities and reactions toward human beings of captive canids match their natural temperament and ecologically adaptive, genetically determined social patterns.

Results

For convenience, the results are broken down into four major sections, and each section is discussed separately. The sections are

Table I.

SUMMARY OF SOCIALIZATION PATTERNS IN CANIDS

Canid type	Primary Socialization	Secondary Socialization	Age of emergence of:			Intrasexual conspecific aggression[a]
			Fear (avoidance)	Aggression (defensive/ offensive)		
Type I (e.g., red fox)	+ (attachment may decrease with maturity)	− (very limited capacity)	4–5 months	4–5 months		++
Type II (e.g., coyote)	+ (attachment to maturity)	± (limited capacity; individual variability)	12 months[b]	24 months		++
Type III (e.g., wolf)	+ (attachment to maturity)	+ (greater capacity; individual variability)	12–18 months[b]	24 months to 4 years		+ (according to age, social rank, etc.)
Domesticated dog	+ (attachment to maturity)	++ (great capacity; breed variability)	± (rare in defective lines)	12–18 months		± (according to age, rank and breed)

[a]Greater in females of all species than in males.

[b]Active submissive approach to strangers after 3–4 months begins to include elements of approach-withdrawal ambivalence in some individuals. This ambivalence increases with age in such animals and by 12 months, withdrawal predominates. Some animals may briefly greet from a distance but are inhibited from approaching.

arbitrarily categorized as follows: permanence of primary social attachments, capacity to form secondary or subsequent attachments, changes with sexual maturity, socio-sexual discriminations. In each section, the behavior of the canids toward human beings is compared and contrasted with their intraspecific behavior in both captivity and in the wild. (See also Table I.)

PERMANENCE OF PRIMARY SOCIAL ATTACHMENTS

All subjects, irrespective of species and sex, remained attached to the author who hand-reared them. They all showed active submissive greeting and passive submission when physical contact was made. The red fox, however, tended to avoid physical contact in spite of showing active submissive greeting and occasional play solicitation. When contact was forced, the fox would remain passive or attempt to escape and occasionally bite. The active greeting by this species involved virtually no physical contact in contrast to the face-oriented licking, pawing, and jumping up of the other canid types. In the wolves, actual contact initiated by the animal, including rubbing and contactual leaning, was of greater duration than in the Type II canids (coyote and jackal).

It may be concluded that in these three canid types, proximity tolerance was lowest in the red fox (Type I), while maintenance of close proximity was greatest in the wolf (Type III). The coyotes and jackals (Type II) were essentially intermediate; in the coyotes, greater individual variation in proximity tolerance developed with increasing age. Some individuals after 2 years of age would only briefly greet, and then withdraw, while at an earlier age would maintain contact and display active and passive submission for a longer period. Others did not show such a change with age. No offensive or defensive aggression was ever displayed by the Type II and III canids toward the author, while such occurrences were not infrequent in the Type I canid. In two instances, the author was subjected to redirected aggression during a conflict between two coyotes; one of the pair redirected its attack on the handler, the bites being inhibited, however. No such experiences have occurred with the wolves. All domesticated hand-raised dogs remained attached through to maturity, some individuals after 5–8 months of age showing a tendency toward passive submission rather than active submission (friendly greeting) when handled by the author. This correlated with a tendency toward avoidance (fear) of strangers in such dogs, but after brief exposure, unlike the wild canids, they would become more active and begin to accept strangers. This capacity for secondary socialization (or social potential per se) contrasts the more limited capacity evident in nondomesticated canids (see below).

CAPACITY FOR SECONDARY OR SUBSEQUENT ATTACHMENTS

Between 1 and 1½ years of age, the wolves began to show increasing wariness of strangers. Prior to this time they would display active and passive submission toward a stranger, but after 18 months of age, an increasing flight tendency was apparent, more so in the male wolf of the pair. Between 2 and 3 years of age, this male became increasingly wary of strangers, but if the person was introduced by the author and remained passive in a squatting position, the male wolf would eventually approach and investigate, vascillating between active and passive submission and flight. The female at this age would usually approach a stranger after a brief period of ambivalence to investigate, greet, and solicit petting. The female was more exploratory even at 3 weeks of age, and at 4 years of age showed more positive social responses and less fear toward strangers than the male.

During the heat period at 2 and 3 years of age, this female showed less passive submission and displayed aggression toward one of the male golden jackals in the opposite cage particularly at these times. After 2 years of age, the male wolf began to threaten certain visitors and at 3 years attempted to attack one student. All agonistic behavior was directed against male human beings, never against women. The possibility that the male wolf is making a sexual discrimination, and upon what basis (possibly olfactory), warrants further study.

A third female wolf recently raised in the same way developed an increasing fear of strange male humans after 5–6 months of age; adult females and preadolescent children were readily accepted even 6 years later. This pattern has been confirmed by others who have hand-raised wolves. What evokes fear of men—odor (androgen metabolites) or human male body langage—remains to be evaluated. With sexual maturity (2 years) this wolf also showed increasing aggression toward conspecific females (and also female domesticated dogs) but displayed active and passive submission toward male wolves and large male dogs. These reactions were more intense during estrus in February.

Of the two male jackals, one was more timid and would only greet and maintain proximity with the author; fear of strangers was evident before 1 month of age in this animal. The other jackal remained friendly toward strangers until approximately 1 year of age. Increasing ambivalence between approaching, greeting, investigating, and soliciting grooming and withdrawing was evident. Subsequently, this animal began to threaten strangers in a highly ambivalent defensive–offensive display, showing more offensive aggression toward men than women. By 2 years of age, he had attacked and slightly injured two females who were regular attendants and one strange female; strange males could not even enter the cage.

The seven coyotes were similarly friendly toward strangers, new, and regular personnel irrespective of sex until after 1 year of age. After this age, all animals began to show increasing hesitation to approach, greet, and investigate strangers, with only one exception. This animal, a female, instead showed increasing aggression toward males and occasionally toward females. One male and two females of the remaining six coyotes became increasingly fearful and would remain in their nest boxes when either regular personnel or strangers came close. The remaining three animals were indifferent, in that they would not greet strangers or regular personnel, but they did not show flight or aggression.

Another male coyote, obtained at 2 years of age provides an additional example. This animal had been hand-raised by two women and remained attached to them as an adult. Around 1½ years of age he was kept and mistreated by a third female and was eventually given to the author by the original owners. This coyote was extremely aggressive toward all females with the exception mentioned earlier. By three years of age, this had generalized to the extent that male human beings were also threatened, but the display showed more ambivalence and elements of defensive aggression, in contrast to the purely offensive display toward females.

Experiments moving coyotes from one enclosure to another and rearranging pairs has revealed a tendency for females to be more aggressive toward strange male and female conspecifics than their male partners. A strange adult male, dominant to the resident female but subordinate to the resident male 6 months after introduction, caused a dramatic change in the pair-bond. The female developed an allegiance with the strange male, who became dominant over her original partner, but in her absence remained subordinate. Any female coyote placed in the cage with these animals would only be attacked by this resident female. The possibility that the dyadic relationship in the coyote is matrifocal (in that the female is the controller and selector of the partner and regulates intraspecific proximity) deserves further analysis in the field.

In the red fox, wariness of strangers was consistently seen as early as 4 months of age, and in only one instance was active submissive greeting displayed toward a person other than that who had been responsible for hand-rearing. This lack of generalization clearly demonstrates the limited capacity of the red fox to develop secondary social relationships.

It should be emphasized that all subjects had more or less the same exposure to people after weaning had been completed between 3 to 4 weeks of age.

CHANGES WITH SEXUAL MATURITY

With the exception of the red fox, many of the above changes in behavior correlate with the attainment of sexual maturity. The Type II and III canids show a gradually increasing tendency to avoid strangers after 4 to 5 months of age, which is clearly evident at 12 to 18 months and tends to persist. In captivity, the female coyotes have shown estrus at 10 months of age, although they do not normally show estrus or breed until the second year of life. The female wolf had her first heat at 2 years, the male showing little interest until the following year. The later emergence of defensive and offensive aggression rather than avoidance reactions toward strangers may be related to this late sexual maturation. In the domestic dog, sexual maturity is attained as early as 6 months of age, but there is a delay in appearance of territorial behavior and aggression toward strangers (both conspecifics and people) until 12 to 18 months of age, and in some cases up to 2 years. This latter behavior may be hormonally dependent since early castration or ovariectomy may prevent such behavior from developing (Brunner, 1968). In the dog therefore, domestication may have caused a split in the timing of gonadal and central nervous system maturation and integration, precocious sexual development being one selected attribute in domestication.

SOCIO-SEXUAL DISCRIMINATIONS

The marked discrimination of human sexuality by some of the Type II and III canids only emerged with increasing sexual maturity. As far as could be ascertained, male assistants did not mistreat the animals and treated them in a comparable way to the females; dress was also often identical, both sexes having long hair and blue jeans! This sex discrimination was most marked in the male wolf and the male golden jackal, overt aggression being displayed toward strange males at increasing frequency after 2 years of age.

The author was attacked by a male wolf during the filming of the "Wolf Men," a television (NBC) documentary. He was a stranger to this wolf, which, together with its mate, had been hand-raised but had infrequent contact with people other than its handler. Prior to releasing the two wolves from their cage for filming in a large enclosure, the female, who was in heat, solicited the author through the cage with full vaginal display and presentation. The male wolf intercepted her, drove her away, and repeatedly threatened the author. When released, the male eventually attacked the author (for further details see Fox, 1971b). This example illustrates clearly one of the problems of socializing wild canids. As a consequence of socialization, they would normally react toward human beings in a manner similar to the way in which they would normally react to conspecifics (this has been reviewed *in extenso* by Fox, 1968c, 1971b); the most logical interpretation of the attack was that the author represented a sexual rival to the male wolf for the attentions of the female. More extreme consequences of socialization are well illustrated in various species of birds (Klinghammer, 1967) and zoo animals (Hediger, 1950) which, when hand-raised, later show a marked sexual preference for human beings. The consequences of socialization with man are less clear-cut in the various canid species. All types studied remained attached to the person who hand-raised them up to their current age of 4 years.* The solitary Type I canid shows a limited capacity for generalization or for the formation of secondary social relationships. This correlates well with the socio-ecology of this Type (Fox, 1975). Several individuals of Type II showed a lesser capacity to form secondary social relationships compared with the wolf. Recent observations of dingos reveal a similar Type II pattern of sociability, a finding verified by Corbett and Newsome's (1975) field study of their social ecology. There is evidence that the Cape hunting dog, another Type III canid, also has a considerable capacity to form secondary social relationships with people after being hand-raised. People who have hand-raised Mexican wolves report that their socialization patterns resemble the coyotes more than the typical wolf. This accords well with the social ecology of this subspecies (*Canis lupus baileyi*) which is closer to the coyote in social behavior, rarely forming packs and more commonly being seen alone or in pairs.

Conclusions

The main conclusions of this study are summarized in Table I. In a review of socialization problems in various nondomesticated, hand-raised carnivores such as ocelots and raccoons (Fox, 1972), a common finding is increasing aggression and unpredictability of behavior toward strangers, and increasing intolerance when disciplined by the owner. The change from the care-dependency relationship of the primary social attachment to a more independent relationship is typical of the relatively solitary carnivores (exemplified by the red fox in this study). Increasing proximity intolerance and parent–infant and infant–infant aggression nor-

---

*Increased aggression and assertion of rank, especially during the breeding season may not be evident in the wolf until 4 to 5 years of age in both males and females.

mally leads to dispersion around 5 months of age in the Type I canids and around 10 months in the Type II canids (Fox, 1975). In the more sociable Type III canid, the care-dependency relationship gives way to a dominance–subordinate relationship combined with allegiance and affection (Fox, 1972). The capacity to form secondary social relationships and to form a stable social hierarchy is limited in the Type II canid and more or less absent in the Type I solitary canid.

Compared to the domestic dog, the wolf seems to have a lesser capacity to form secondary social relationships. There may be a period after which such relationships are difficult to establish in the wolf. This may be correlated with the fact that the wolf pack, especially the upper echelon of mature dominant wolves, is closed to strange wolves. Strangers are avoided or driven away, while lower ranking yearling wolves will more readily accept strangers and may even leave the main pack (Fox, 1973). The social behavior of the more mature wolves is an ecologically adaptive social mechanism that regulates pack size. In the domestic dog, this behavior may have been selectively eliminated in many breeds so that as adults they will readily accept strangers. The possibility of infantilism (Zimen, 1970) or neotenic perseverance of infantile care-soliciting and submissive behavior in the domestic dog may facilitate the generalization of secondary socialization as well as increasing proximity tolerance. Most significant, however, is the evidence for innate social predispositions which may facilitate the domestication of some species but be a major limiting factor in others of different temperament and capacity to develop secondary social relationships.

Summary

The social consequences of hand-rearing wild canids in terms of their actions to their foster parent and to other human beings have been reviewed. The canids studied represent the three socio-ecological types, namely the solitary red fox, gregarious wolf, and intermediate coyote and golden jackal. Differences in permanence of primary social attachments, capacity to form secondary social relationships, changes with sexual maturity and socio-sexual discriminations are discussed. The various relationships between socialized canid and man reflect the innate social capacities of the species in question and also correlate with the socio-ecological patterns of the species under natural conditions. In other words, it is shown that the socio-ecologically adapted temperament and social capacities of each canid type are primary determinants of their reactions toward human beings after early socialization with man. Where relevant, the social behavior of domesticated dogs is compared and contrasted with that of the wild species; in contrast to the dingo and wolf (proposed ancestors of *C. familiaris*), most breeds of dog have a greater social potential or ability to establish secondary social relationships which, in contrast to these other species, must be a consequence of domestication per se.

IX

Stages and Periods in Development: Environmental Influences and Domestication

Introduction

The purpose of this chapter is to review a number of studies that serve to illustrate various principles of neural and behavioral development. An attempt is made to bring together several diverse topics from the ethological and psychological literature in order to formulate some general concepts pertaining to the analysis and interpretation of critical and sensitive periods in development. These periods represent developmental stages at which environmental (i.e., human) influences can have profound and long-lasting effects upon later behavior, physiology, and emotional reactivity. A number of terms in current use that attempt to describe some of these developmental phenomena are reviewed, principally to clarify the use of such terminology in this somewhat diffuse area of developmental psychobiology. Structural and functional development of the canine brain is correlated with certain aspects of behavioral ontogeny underlying the critical period of socialization.

The degree of plasticity and adaptability of the developing brain and behavior can be evaluated, for example, by environmental manipulations such as sensory deprivation, social isolation, or excessive stimulation (handling, experiential enrichment) at different ages or stages of development. These various treatments that may modify brain and behavior development are considered in relation to animal domestication.

Structural and Functional Development of the Canine Brain

Myelinization of the canine central nervous system (CNS) occurs in two cycles. At birth and during the first 3 weeks, there is a gradual increase in myelin content of the spinal cord occurring earliest in the cervical region and in motor roots prior to sensory roots. The corticospinal tract is the last major efferent pathway to develop. Subcortical structures posterior to the posterior commissure also begin to myelinate during this early postnatal period (Fox, 1971a). Cranial nerves associated with feeding and cephalic cutaneous sensitivity (facial and trigeminal nerves) and with balance and body-righting (nonacoustic portion of the 8th cranial nerve) are well myelinated at birth. Between 3 and 4 weeks of age, the second cycle of myelinization occurs, anterior to the posterior commissure; specific and nonspecific thalamocortical afferent fibers become myelinated (Fox, 1971a). The various regions of the neocortex do not show an equal increase in myelinization. Myelin is first detected in the somatosensory area at 4 weeks and by 6 weeks of age is more evenly distributed in other regions such as the visual and auditory cortex. The frontal lobe shows the most gradual myelinization.

In contrast to this more gradual myelinzation of the neocortex, neuronal development in terms of cell density, neuronal size, and apical and basilar dendritic complexity, attains relative maturity in the sensorimotor, visual, and auditory cortex by 6 weeks of age.

Also, the size and elaboration of the processes of fibrous astrocytes closely resemble the adult between 5 and 6 weeks of age (Fox, 1971b).

From these observations, it is apparent that neocortical neuronal maturation precedes myelinization and that the second myelinogenetic cycle does not begin until neuronal and glial development is well advanced.

Detailed qualitative studies on the postnatal development of the canine EEG during wakefulness, quiet sleep, and paradoxical (REM or activated) sleep show that the first signs of slow wave (possibly thalamocortical) activity appears around 17 days, and by 4 weeks of age, a clear distinction between various states of wakefulness and sleep can be detected on the EEG (Fox, 1971b). Relatively mature patterns of electrocortical activity are present at 5 weeks of age, at which time the percentage of various states of sleep and wakefulness are relatively mature. Only slight qualitative changes in EEG occur after 5 weeks of age, notably the preponderance of fast spindle activity at the onset of quiet sleep, a gradual reduction in amplitude and increase in fast frequency components during wakefulness, and a more gradual increase in amplitude during drowsiness and quiet sleep. In close temporal association with EEG development, auditory and visual evoked potentials, in terms of latency, attain mature characteristics between 5 and 6 weeks of age.

Behavior Development in the Dog

Several parameters of behavioral development in this species have been investigated and correlate well with the above data on CNS development. The dog is neurologically mature at 4 weeks of age with the exception of equilibration and adult locomotor abilities such as running and leaping (Fox, 1971b). Between 3 and 4 weeks of age the pup begins to interact with its socioenvironmental milieu and to develop primary social relationships or emotional attachments. This marks the onset of the critical period of socialization

(Scott and Fuller, 1965), and the neurological and behavioral events are briefly reviewed since they are important factors which underlie the beginning of the critical period.

The data on the developing canine brain serve to demonstrate the temporal coincidence of development and maturation of several interrelated structural and functional parameters. This coincidence, which occurs between the fourth and fifth postnatal week in the dog, may be termed a period of integration. It is at this time that the several parts of the developing nervous system show both structural and functional integration, which marks the beginning of a relatively mature organizational level of activity.

At this time, the organism begins to interact rather than react with conspecifics and through social experiences with both parent and peers develops emotional attachments to its own kind or to man. If denied human contact during this critical period from 4–12 weeks of age, it will subsequently avoid human contact (Scott and Fuller, 1965). Such dogs are fearful of humans and are virtually untrainable. The fear period which develops after 8 weeks of age limits the capacity to develop new social attachments and essentially terminates this critical socialization period. Thus even in a domesticated species, lack of exposure to man during this formative period (when brain centers are integrating and emotional reactions developing) will greatly limit the social potentials of the species.

Imprinting and Socialization

The concept of a critical period implies that experience at a particular time of development is essential for normal development to continue. This is exemplified by the phenomenon of imprinting in birds. Imprinting means attachment, and birds (such as ducklings) that are relatively mature when hatched normally become imprinted onto the mother during the first few hours after hatching. If they are taken as soon as they are hatched and are raised not with the mother but with a human being, a flashing red light, or a moving cardboard box, they become preferentially attached to the species or the object with which they have been raised. Some reversal is possible during subsequent weeks, but there is strong evidence that this early exposure results in a very specific and enduring attachment—so enduring that at maturity social and sexual behavior may be directed toward the same stimulus to which the bird was imprinted early in life.

Lorenz (1970) hand-raised jackdaws and crows and found that when they reached maturity they would show courtship behavior toward him and would attempt to mount his hand. During the time when they would normally be taking care of their own young, they would attempt to stuff grubs into his ears. Hediger (1950) describes the experience of one of his zoo keepers who hand-raised a male moose. When the moose reached sexual maturity, the keeper led it into a field of female moose, and the young bull became sexually aroused. Instead of directing his sexual behavior toward a female moose, however, he attempted to mount his keeper. Such bizarre behaviors are good examples of the enduring effect of imprinting in determining later social and sexual preferences in various animals. Klinghammer (1967) has shown that the effects of hand-rearing in various species of pigeons can vary. One species, for example, if hand-raised, shows a sexual preference at maturity exclusively for its human handler rather than for its own species. Other closely related species show a reversal at maturity; although still friendly toward the human foster parent, they are only sexually attracted to their own species. Klinghammer also identifies a third category of pigeons in which individuals have the best of both worlds and show sexual behavior toward both their own species and the foster parent!

It is also known that the two sexes of a given species are not affected in the same way by imprinting. In mallards, for example, Schutz (1965) finds that early imprinting in the males later determines their sexual preference while female mallards, regardless of imprinting, show an innate preference for males of their own species. This was confirmed by raising male and female mallards with different species of ducks. When they reached maturity, the male mallards preferred the species with which they had been cross-fostered while the female mallards tended to reverse their social preference in sexual encounters and to seek out males of their own species.

Dogs have been raised with cats during the critical period of socialisation to evaluate further the effects of cross-fostering. In this study a 3-week-old Chihuahua was placed with a litter of 4-week-old kittens. Five replications of the study were done (Fox, 1971a), and in each replication a battery of tests was given to the pup at 12 weeks of age. It was found that the pups raised with kittens made no social responses to their own mirror images; they literally lacked species recognition. They also preferred the company of cats to that of the littermate Chihuahuas which were used as controls. The cats that had been raised with a dog were also sociable toward dogs that had not been raised with cats. In contrast, cats that had had no earlier exposure to dogs avoided contact with them—with one exception. Salzen and Cornell (1968) did a comparable study for color preferences in chicks, which is reminscent of the role integrated schools might play in achieving interracial socialization. They raised a green-dyed chicken with a group of red-dyed chickens. When they placed these chickens together with a new group of chickens that were all dyed green, the green chicken ran toward his red companions and did not mix with the green group. Variations on this experiment seem to confirm that allegiances and social preferences are based on early social learning.

Even more subtle consequences of early rearing can affect social preferences. One experiment used three groups of pups. The pups in one group were hand-raised and were exposed only to humans. The pups in the second group were weaned early and had almost equal contact with humans and dogs up to the point of testing. The pups in the third group were raised with each other (Fox, 1971a). When the pups were placed in new social groups, it was found that they tended to segregate themselves in accordance with their rearing history. The pups that had been hand-raised and had no social experience with their own kind tended to stay together while the pups that were weaned early and the controls tended to segregate into like groups. Other aspects of social behavior were affected also.

Sackett et al. (1965) found a very similar social consequence of early rearing experiences in young rhesus monkeys. They too used three groups. One group was hand-raised exclusively with people. The second group had contact only with other monkeys. The third group had more or less equal contact with humans and their own kind. The monkeys segregated into like groups for social play during early life and later clearly demonstrated sexual preferences based upon their rearing experience.

If social experiences are denied during the early critical period, as demonstrated in our cat–dog socialization study, there is often impairment in the subsequent development of social relationships. Scott and Fuller (1965) showed that if dogs are denied human contact until approximately 14 weeks of age, they are wild and unapproachable. Scott and Fuller demonstrated clearly the critical nature of the socialization process and found that human contact between 6 and 8 weeks of age seems to be optimal for the socialization of dogs (see Figure 1).

Although there is an optimal period for socializing pups, there is evidence that dogs may subsequently regress or become feral. The social bond with man may be broken when well-socialized pups are placed in kennels at 3 or 4 months of age; by 6 or 8 months they are shy of strangers and often of their caretakers if they have not been handled much. In addition, they may be extremely fearful when removed from their usual quarters. Their fearfulness is the result of a combination of institutionalization and desocialization. In connection with the phenomenon of desocialization, Woolpy (1968) found that captive wild adult wolves can be socialized in 6 months of careful handling and that when they are subsequently given less human contact they do not regress or become desocialized. In contrast, wolves that are socialized early in life are like dogs in that they will regress if they are subsequently given less human contact. These findings suggest, therefore, that although there is an optimal period early in life when socialization can be rapidly established, subsequent reinforcement is necessary because of some intrinsic instability of retention in young animals. Woolpy (1968) concluded that

an important aspect of socialization is learning to cope with a previously unfamiliar environmental situation in the presence of unreduced subjective fear.... We have interpreted the results of both the tranquilizer and the socialization experiments

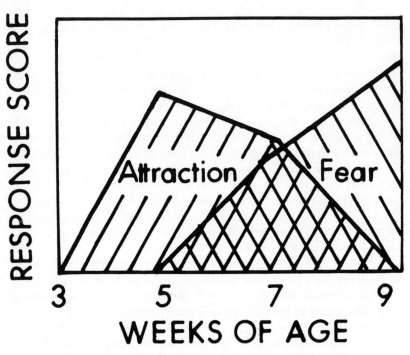

Figure 1. *Socialization in the dog involves an initial attraction phase, but if denied human contact until first tested at 5, 7, or 9 weeks of age, fear and the tendency to flee increases with increasing age. The optimal period for socialization is between 6 and 8 weeks. (Adapted from Scott and Fuller, 1965.)*

to indicate that the fear of the unfamiliar is the primary obstacle to wolf socialization and that, while the overt responses of fear appear very early in life, its subjective components continue to develop throughout at least the first year. Socialization must be conditioned in the presence of the fully developed subjective components of fear, and hence it cannot be permanently maintained in juveniles if they are left to develop fear responses subsequent to having become socialized in early life.

There seems to be a comparable critical period for the development of emotional attachments in children. Bowlby (1971), from the Tavistock Clinic in London, has placed much emphasis upon the critical nature of infant socialization in determining later socially adjustive behavior and in preventing delinquency and even antisocial and sociopathic behaviors. It seems that in both animals and humans the initial or primary socialization early in life is the basis for the development of subsequent secondary social relationships; if primary attachments are not made or are in some way modified, the consequences for the later social adjustment of the individual can be quite serious, and in terms of domestication greatly limits the potentials of the animal for human use.

Socialization and Overattachment

The "perpetual puppy" syndrome may develop in adult dogs if a symbiotic relationship has been maintained by overindulgent and permissive owners, and severe anaclitic depression may follow separation of the dogs from their owners for surgery, boarding, or quarantine (Fox, 1968d). It should be emphasized that as a consequence of the symbiotic relationship with the owner the pet may develop a variety of care-soliciting (et-epimeletic) reactions such as whining, jumping up, following constantly throughout the house, crying when left alone, and submissive urination. These symptoms resemble regression in man to more neotenic or infantile behaviors. Punishment after the disturbed pet (or child) has urinated, defecated, or generally pestered the owner sufficiently may lead to a masochistic form of reinforcement.

"Sympathy" lameness, hysterical paraplegia, and coxalgia have been described in dogs (Fox, 1968d) and are well documented in man as attention-seeking reactions. Chess (1969) stresses the correlation between acute illness and dependency behavior in human infants. After recovery, the child attempts to maintain the interpersonal relationship that brought him special attention during the illness. Some overindulged pets have been known to refuse to use one limb after surgery for congenital patella luxation because they received so much attention and petting from their owners while they were recovering. In extreme cases, muscle atrophy developed; in others, surgical recovery was complete, but the subjects would suddenly become lame when they were in an anxiety or conflict-provoking situation and were soliciting the attention of their owners.

Environmental and Experiential Influences

Environmental and experiential factors which influence development are reviewed briefly to provide a basis for further discussion of the importance of environmental influences in the process of animal domestication.

Schneirla (1965) has developed a very important theory involving the significance of approach–withdrawal processes in the organization and development of behavior. For instance, approach and subsequent reward (such as contact comfort) are tied in with parasympathetic arousal, while withdrawal from painful stimulation is associated with adrenal-sympathetic arousal. Contact comfort, especially that associated with nursing or petting an animal, causes parasympathetic arousal (exemplified in the human infant by salivation, increased peristalsis, secretion of digestive juices, general relaxation, and occasionally penile erection). Young animals derive considerable reward from contact comfort, grooming, and nursing; if the same circumstances cause parasympathetic arousal, then contact comfort, grooming, and nursing would tend to improve digestion and assimilation of food as well as to facilitate emotional attachment (Fox, 1975b). This "gentling" or petting phenomenon is part of the taming-domestication process. Even gentling a pregnant animal can result in offspring that are more docile, and as emphasized by Denenberg and Whimbey (1963), this may be a significant fact in the domestication process.

Several years ago Spitz (1949) demonstrated that inadequate mothering, which may be reinterpreted as inadequate parasympathetic arousal, led to a wasting disease in many orphan children.

These children did not gain weight, did not adequately digest and assimilate their food, and many succumbed to infections. When Spitz initiated a regime of mothering, the infants began to gain weight, and the rate of mortality decreased significantly. This effect, distinct from the handling effect, has been termed gentling. Bleicher (personal communication) has found similar effects in handled and nonhandled orphan pups raised in social isolation.* No hormone has yet been identified, but it is possible that the gentling of pregnant rats influences the parasympathetic nervous system and that certain neurohormones affect the developing fetus.

A number of independent studies on rodents supports Schneirla's theory, demonstrating that experimental manipulations early in life can have long-lasting effects on later behavior. For instance, if a 5-day-old rat is taken out of its nest and is placed in a metal pot at room temperature for 3 min a day for 5 days and then is allowed to mature, as an adult it is less emotional than littermates not treated in this way. It may also be more resistent to physical stress such as terminal starvation, cold exposure, and certain pathogens. This effect, which has been called the early handling effect, seems to influence the adrenal–pituitary axis of the developing rodent. Denenberg (1967), Levine and Mullins (1966), and other workers have studied this phenomenon in great detail. It would appear that the stress to which the neonate is exposed in some way affects the way in which it responds to psychological and physical stresses later in life. It has been proposed that a hormonostat exists in the neurohypophysis which is "tuned" to the adrenal glands during the sensitive period when the rat is between 5 and 10 days of age. When this is tuned experimentally by a sudden elevation of corticosteroids during the sensitive period, the hormonostat operates differently when the organism is stressed in maturity. Typically, as Levine has demonstrated, the stress response in adult control rats is somewhat maladaptive. There is a long latency period before the neuroendocrine system responds, but the eventual response can be long-lasting and at times may

---

*These effects have also been found in calves (Professor C. Schwabe, personal communication). Mortality rates were lower on farms where calves were handled more often and with more care and affection.

trigger the onset of Selye's stress syndrome. Rats handled early have a much shorter latency of response, and the duration of response is shorter. In effect, their reactions are not unduly delayed and they do not overreact.

In some strains of mice that develop spontaneous leukemia, the onset of the condition can be delayed by this handling procedure. Handled mice also have a greater resistance than nonhandled controls to implanted tumors. There are also genetic or strain differences, as demonstrated by Ginsburg (1968). For one strain, a given level of stimulation might be excessive; in another strain, the sensitive period might lie between postnatal days 6 and 10 or between postnatal days 10 and 15. Ginsburg emphasizes that in a typically heterogeneous population, as in *Homo sapiens*, there would be a normal distribution curve and that individuals would have different sensitive periods as well as different response thresholds. In view of these individual and genetic variations, therefore, we must be very careful in making generalizations about the handling effect.

Levine points out that in the relatively "overswaddled" laboratory environment neonate rats may well be understressed and that the handling procedure is much closer to the kind of experience they would normally have in the wild. Under natural conditions, the mother frequently leaves the nest to forage for food, and there are various environmental changes, all of which could have an additive effect resulting in an adult animal that is psychophysiologically superior to the "overswaddled" laboratory specimen.

In a study of the handling phenomenon in dogs (Fox, 1971a), pups were subjected to varied stimulation—exposure to cold, vestibular stimulation on a tilting board, exposure to flashing lights, and auditory stimulation—from birth until 5 weeks of age. The pups in the study differed from the controls in a number of ways including earlier maturation of EEG, lowered emotionality which enhanced problem-solving ability in novel situations, and dominance over controls in competitive situations. Analysis of their adrenal glands indicated a fivefold increase in norepinephrine, and studies of their heart rates indicated that a greater sympathetic tone was developed as a consequence of the early handling stress. Meier (1961) demonstrated comparable maturation of EEG and behavioral changes in superstimulated Siamese kittens. (See Chapter 10.)

A generalization cannot be made about optimal handling for a given species. For instance, it appears possible that a given level of early stimulation may produce psychophysiological superiority in some human infants and pathophysiological inferiority in certain other individuals. As Thomas *et al.* (1970) have demonstrated in their longitudinal studies of human infants, handling has to be very carefully adjusted to the basic temperament and autonomic tuning of the individual. The effects of early handling appear to be on the adrenal–sympathetic system, and the research evidence suggests that early handling not only resets the pituitary–adrenal axis but in some way influences autonomic tuning and temperament or emotionality.

To what extent current animal husbandry practices and various handling regimes adopted in early domestication affect farm livestock and house pets alike, remains to be evaluated. Those breeders who have adopted the above handling stress program for puppies and also the U.S. Army Veterinary Corps (Biosensor Research Division) report extremely promising results in terms of later stress—resistance, emotional stability, and improved learning ability. This phenomenon may represent a new tool in animal husbandry and domestication.

Prenatal Influences

Prenatal anxiety can have a significant effect on the emotionality and the later personality developed of an infant (Joffe, 1969). Thompson (1957) was one of the first to demonstrate the effects of prenatal anxiety on the behavior of offspring. He developed a conditioned emotional reaction in rats placed in a shuttle box; whenever a bell rang, the rats had to jump to the safe side of the box in order to avoid shock. The basis for an emotional reaction was established by placing a barrier in the center of the box so that the animals could not jump to the safe side when the bell rang. Since the rats were never given shock when the bell rang after the barrier was erected, the reaction was purely emotional. The offspring of pregnant rats that were stressed in this way were cross-fostered by normal mothers in order to control for any postnatal

transfer effects from the mother. Thompson found that the prenatally stressed rats were much more inferior because heightened
emotionality interfered with their performance. He was able to
produce a similar effect by injecting adrenalin and ACTH into pregnant females.

Joffe (1969) has reviewed many such experiments including his
own on the Maudsley reactive and nonreactive strains of rats. He
was able to demonstrate genetic differences in susceptibility to
prenatal stress as well as differences correlated with the sex of the
offspring and the period during pregnancy when the stress was
administered. He discovered that premating stress could also have
a significant effect on the behavior and the reactivity of the offspring. These experiments emphasize that the developing
phenotype is modifiable not only after birth but also prenatally, the
effect being mediated by the neuroendocrine system.

These prenatal and postnatal studies in modifying the
phenotype also show that relatively little is known about how to
provide environmental stimulation and rearing programs which
will ensure that the developing phenotype can be optimized or at
least how it can be protected from deleterious influences in food-
producing farm animals. There is much discussion about genetic
engineering and programmed breeding through artificial insemination, but the effects of environmental phenomena that are present
during certain critical periods in development have barely begun to
be explored.

Environmental Enrichment and Deprivation

In *The Descent of Man*, Darwin observed that the domesticated rabbit has a much smaller brain than its wild counterpart and attributed this to the cumulative effect of generations of captivity during
which each generation received far less stimulation than it would
have in the wild. To test the effect of environmental stimulation on
captive animals, Bennett and his co-workers (1964) at the Lawrence

Radiation Laboratory in California raised rats in enriched environments. Typical laboratory rats were placed in a large cage with cage mates and with various objects (wheels, runways, and so on) to manipulate. Under these conditions of environmental complexity, Bennett found that the rats in the study became more exploratory than the controls and that they developed significant differences in brain size, in the depth of the visual cortex, and also in the turnover rates of brain acetylcholinesterase. More recently Bennett demonstrated that the enrichment effect is dependent upon an interaction between the inanimate environment and conspecifics. That is, if a rat is raised alone in an enriched environment, profound changes in brain and behavior do not occur; nor do they occur if a rat is raised with companions in a barren cage.

Dogs have been raised under a schedule of paced increments of experience, periodically taking some of them out of their home cages and letting them explore an arena containing novel stimuli (Fox, 1971a). Those that were allowed in the arena at 5, 8, 12, and 16 weeks of age for a mere ½ hr per exposure explored increasingly as they grew older and developed a preference for more complex stimuli as they matured. Littermates that were placed in the arena for the first time at 12 or 16 weeks of age did not explore; they withdrew or did not leave the start chamber, and many of them were catatonic with fear. We are dealing here with an institutionalization syndrome; those dogs that did not have an opportunity to leave their home cages until sometime after 8 weeks of age could not tolerate the complexity of the environment and so they withdrew to avoid overstimulation. (Later some of the neurological effects of the emergence from isolation are discussed.) Sackett (1968) adds support to our canid study in an experiment that he did with rhesus monkeys. He found that as rhesus monkeys get older, they prefer to look at picture cards of increasing complexity. If monkeys are raised in isolation, at 6 months of age they prefer to look at cards of less complexity than those which rhesus monkeys normally prefer at that age. He concludes from these experiments that in the absence of paced increments of experience the organism will seek a lower level of stimulation and environmental complexity.

From these studies an arousal-maintenance model of perceptual-motor homeostasis can be formulated wherein the optimal arousal or tolerance level is set early in life as a result of the quality and quantity of early experiences. The level is set low in those animals that have had few increments of experience. If the environment does not provide varied stimulation, the subject may compensate by creating its own varied input by elaborating stereotyped motor acts or by directing specific activities toward inappropriate objects (such as copulating with its food bowl). The stereotyped motor acts (thumb-sucking, self-clutching, and rocking in primates) developed while in isolation may be performed when the subject is in a novel environment and may serve to reduce arousal or anxiety because they are familiar activities and may be comforting (Berkson, 1968). This type of stereotype is to be distinguished from the locomotor cage stereotypes described by Meyer-Holzapfel (1968) which are derived from thwarted attempts to escape. Mason (1967) has developed a comparable theory in which he uses the term "general motivational state" in reference to the influence of the degree of arousal on the organization of behavior during different periods of development.

In reviewing their own studies of primates, Jensen and Bobbitt (1968) show how the deprivation of inanimate play objects affects social behavior.

> Inanimate objects constitute another important class of environmental factors. Our tests of the effects of enrichment by toys indicate that lack of them and climbing facilities will seriously handicap the young animal in learning motor skills, in developing independence from its mother, and in interacting with peers. Further, these effects may be critical in determining later social dominance. We hypothesized that prior and concurrent toy experience facilitate peer socialisation.

> In an impoverished environment such as a barren cage, there is a prolonged period of mother–infant closeness which is essentially a retardation in the mutual independence process. Mother–infant pairs in an enriched environment manipulate themselves and each other less and manipulate the environment more. Jensen and Bobbitt (1968) propose a continuous process of detachment of infant from mother (and vice versa) as attachments to the larger environment develop. Apparently deprived infants at 6 months of age are severely handicapped in social responsiveness when faced with an enriched peer, with respect to whom they are subordinate.

These authors emphasize that short-term maternal separation or isolation from an enriched environment may actually intensify interaction when the infant is returned. This temporary increase in responsiveness after a brief period of separation (in effect, an instance where absence makes the heart grow fonder) clearly shows the importance of the duration of deprivation in addition to the type of deprivation and the age of the organism when deprivation occurs.

As in all developmental problems, the role of genetic factors must be considered first. Henderson (1970) has demonstrated elegantly the interaction between genetic and environmental influences in the development of mice. Working with several strains, he found that the results with mice raised in enriched environments were somewhat similar to those obtained by Bennett and his colleagues (1964). The mice were superior to the controls raised under standard laboratory conditions in performance tests including various learning tasks and motivational tests of exploratory behavior. Henderson proposes that there is an environmental repression op-

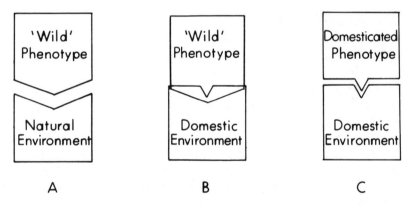

A   B   C

Figure 2. *The domesticated phenotype ultimately (and ideally) fits with the type of environment in which the animal is kept (i.e., analogous to a process of pseudospeciation where different genotypes and phenotypes in the same species may arise as a consequence of different domestic environments. The fitness transitions between (A) and (B) and (B) and (C) may be critical in terms of behavioral adaptation.*

erating as a consequence of rearing under the relatively im-
poverished environment of the laboratory. He has also demonstrated that hybrids of the various mouse strains raised under
standard laboratory conditions performed much better than the
pure parent strains in various tests given to them at maturity (a
typical example of hybrid vigor). When Henderson raised hybrids
in enriched environments, their performance scores were also
higher than those of pure strains that were raised under the same
conditions. Henderson's paper is an important one for everyone
working with laboratory animals. The extent to which the various
artificial environments in which domesticated animals have been
raised for generations (e.g., home environment for cats and dogs
and "intensive" systems for pigs, poultry, and calves) have influenced brain and behavior remains to be evaluated (see Figure 2).

Socioenvironmental Influences and Reproduction

Although imprinting and socialization influence sexual preferences
in later life, other factors relevant to domestication and present
husbandry practices must be considered. Vandenberg (1969), for
example, has shown that the presence of a mature male mouse
accelerates sexual maturation in females compared to females
raised in a unisexed group. There is also strong evidence from wild
animals raised in zoos that being raised together per se (in
heterosexual pairs) may in some way inhibit reproduction. This has
been confirmed experimentally by Hill (1974) who found that pre-
pubertal familiarity in deer mice delays reproduction in nonsibling
pairs. Such a reproductive delay, he proposes, may act to reduce
inbreeding depression and regulate population growth. A third
phenomenon, the social facilitation of reproduction (by visual, auditory, or olfactory cues, depending upon the species) known as
the Darling effect, is pertinent to this review. Animals in social
isolation or in small social groups have a lower reproductive effi-

ciency than those in larger groups. A male (e.g., a bull) will have a higher sperm count if given access to females compared to one that is socially deprived; while females together (e.g., rodents and dogs) may show a synchronization of estrus.*

These social and developmental influences on reproductive physiology and sexual behavior warrant further study in domesticated animals and serve to emphasize the close relationship between neuroendocrine activity and the social environment.

Sensitive and Critical Periods

Some of the complexities of ontogeny and integrative activity of the brain and behavior are being opened up for future investigation through these studies that determine the effects of environmental influences on certain aspects of neurobehavioral development. Within certain limits, environmental influences can modify genetically predetermined patterns of development and modify the later maturing phenotype. The type and intensity of environmental stimulation may have different effects at different ages (Denenberg, 1964) and at different ages in different strains (Ginsburg, 1968). Similarly, environmental and experiential deprivation may produce very different effects at various ages (Scott, 1962; Fox and Spencer, 1969). The integration of structuro-functional components within and associations between different systems may also be affected (Riesen, 1961) as well as producing nonspecific or more general effects such as influencing the general level of arousal, emotionality, and response threshold to noxious or novel stimuli (Denenberg, 1964). Modification of the phenotype following environmental or experiential manipulations, an *effect which may be transmitted through successive generations* (Denenberg and Rosenberg, 1967), may result in psychophysiological superiority or pathodevelopmental abnormalities, depending upon the intensity

*Possibly large numbers of animals caged in close proximity may show a decrease in fertility—a phenomenon suspect in large breeding kennels.

of stimulation or deprivation and also the age and strain of the species used.

Isolation, deprivation, or selective stimulation and enrichment may alter the normal ontogenetic sequence of behavior development so that certain patterns may disappear earlier with maturity or persist for an abnormally long time. Meyer-Holzapfel (1968) has shown that with continued reinforcement, the gape response of birds may persist and not disappear with maturity at the normal time. Later developing patterns may not appear at the normal time because of inadequate or conflicting stimulation or reinforcement, as shown, for example, by Kovach and Kling (1967) in their studies of feeding behavior of kittens fed by stomach tube and deprived at various ages from nursing. To evaluate normal ontogenetic processes by such techniques may be misleading; what might be shown is the degree of modifiability or adaptability of the behavioral phenotype within certain sensitive periods and the dependence (or independence) of development and integration upon stimulation, reinforcement, and paced increments of experience.

A distinction is now drawn between critical and sensitive periods because these terms have been used indiscriminately in the past to describe temporally discrete periods in an organism's lifetime when a number of very different experiences such as handling (Denenberg, 1964), social experiences (Scott, 1962), and imprinting stimuli (Bateson, 1969) have their greatest and most enduring influence. This distinction is necessary because some experiences may be essential for development to continue normally (i.e., experientially dependent development); whereas other behaviors and organismic functions may not be dependent upon such influences for their normal development, yet may be affected at a particular time early in life by these environmental influences (i.e., development is experientially independent but modifiable). A third category of behaviors and functions may be experientially independent and essentially unmodifiable: these are discussed subsequently (see Figure 3).

Critical periods may be regarded as those definable times during development when the organism is dependent upon environmental influences for its development to continue normally. Such experiential dependence at one critical period may represent an epigenetic crisis (Fox, 1970b; King, 1968), the onset and duration of

MATURATION

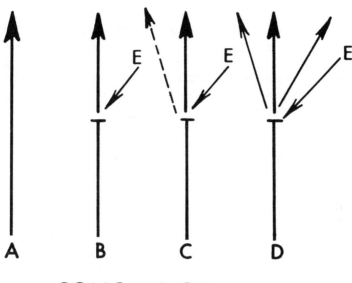

CONCEPTION

Figure 3. *Highly simplified schema of various patterns of matura-*
tion where environmental influences (E) may or may not have a
determining influence, as exemplified by various human interven-
tions in domesticated animals. A, environmental independent/
resistant development, e.g., of barking in dogs and of "fixed" dis-
plays. B, environmental dependent/expectation where (E) occurs
during a critical period, providing input necessary for development
such as for song in some birds. C, direction of development may be
modified by input during a sensitive period early in life (but normal
maturation is not dependent upon this input) e.g., early handling
stress. D, varying environmental input during such a sensitive period
may influence maturation in many directions, resulting in considerable
phenodeviance from the norm. (From Fox, 1975b)

the period being species-characteristic and therefore genetically programmed. There may also be interdependence between critical periods, where one critical period has an induction effect on a later period. One example of such delayed effects between experiences during an early critical period and a period later in life, when behavior is activated and is influenced by experiences during the critical period earlier in life, is the development of song in certain birds (Nottebohm, 1970). Another example is the early critical period for primary socialization which subsequently influences the development of secondary social relationships (Scott, 1962).

Sensitive periods differ from critical periods in that the organism is not dependent upon stimulation during such periods for development to proceed normally. It is rather that at particular periods in development the organism is especially vulnerable to certain environmental influences (temperature changes, electroshock, and experimental hormone manipulations) (Morton, 1968) which may have profound effects on subsequent development. Experimental manipulations during such sensitive periods therefore may influence the integration of component parts of a particular system, e.g., adrenal or gonadal–hypophyseal axis (Levine and Mullins, 1966), and have long-lasting effects which can be detected later in life.

In summary, therefore, critical periods may be regarded as those times in early life when the organism is developmentally dependent upon certain exogenous stimuli which are normally present in its socioenvironmental milieu. Sensitive periods, in contrast, are those times when the organism is especially vulnerable to environmental changes (change in temperature, maternal deprivation), the consequences of which are long-lasting.

The developing organismic system is a link in the intrinsic complexity of interrelated systems extending in two dimensions; one in time, between ancestors and future generations and the other, in space or place—the organismic system within a complex of ecological and social systems. The investigator may qualitatively catalogue, quantify, and dissect development at one of two levels, at the level of the process itself, or at the level of the interrelationships between processes or systems. Eventually he must reconstitute reality; that is, he must describe the developing organism in relation to its environment. Then developmental changes in terms of genetically programmed, ecologically tuned or adaptive expectancies, or human-induced changes via domestication may be identified.

The potential environmental influences (as distinct from genetic selection per se), subsumed under the general phenomenon designated loosely as domestication, which may greatly alter a given species over successive generations are reviewed in this chapter. Early handling stress (or prenatal gentling), imprinting and socialization, and environmental enrichment (or deprivation) during sensitive and critical periods in early life represent the developmentally timed stages when appropriate input may dramatically influence development and later behavior. This is illustrated schematically in Figure 4 and represents the three key interfaces

Developing and Integrating Nervous System

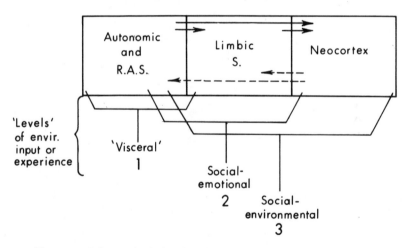

Figure 4. *Schema of relationships between stage of development of the central nervous system and the times when various types of environmental influence have their greatest effect. 1, Handling early in life affects the adrenal–pituitary, autonomic, and reticular activating (R.A.S.) systems. 2, Development of social attachments correlates with the development of limbic (emotional) centers. 3. Social and environmental influences affect brain maturation.*

between the developing organism and its environment where human domesticating interventions may have their greatest effect.

Emotional reactivity is reflected in autonomic changes in cardiac activity. This most sensitive indicator has been utilized in a series of studies to evaluate some of the variables (such as the effects of early handling stress) discussed in this chapter. A major consequence of selection for docility in animal domestication may influence autonomic "tuning" and such possibilities will be investigated in Chapter 10.

X

Behavior, Development, and Psychopathology of Cardiac Activity in the Dog

Introduction

Weber and Weber (1845) first demonstrated that stimulation of the vagal nerve causes cardiac inhibition. More recently, Long *et al.* (1958) reported that after cutting the vagal nerve, the postoperative rate in dogs was almost doubled due to the absence of vagal inhibition.

Belkin (1968) showed that this vagal inhibition may be an important behavioral mechanism in many vertebrates. He demonstrated bradycardia associated with fear, in lizards, nestling birds, hares, turtles, and man. Hofer (1970) has completed a detailed study of this "fear bradycardia" during sudden and prolonged immobility in rodents. Various cardiac arrhythmias and polypnoea were recorded, and in one species, a torporlike reaction, rather than tonic immobility, was observed. Richter (1957) observed fatal cardiac arrest apparently mediated by the parasympathetic system, in wild Norway rats handled and tested in the laboratory.

Bradycardia is also seen in the diving reflex in many diving birds, in seals, and in man (Scholander et al., 1942). Gellhorn, in elaborating a useful theory on autonomic tuning of the central nervous system, observes that

The immersion of the face in cold water leads to a vagal reflex which increases the resistance of the organism to asphyxia and is therefore of great biological importance in diving animals. Studies on this reflex in man with recordings of blood pressure, heart rate, oxygen saturation and skin temperature disclose that the immersion evokes an immediate slowing of the heart rate which persists throughout the test. A rise in blood pressure and cutaneous vasoconstriction appear after the bradycardia. This reflex is therefore an example of a (parasympathetic) trophotropic reaction followed by compensatory ergotropic (sympathetic) phenomena. If the immersion test is carried out on an "apprehensive" subject the vagal reflex is greatly increased, illustrating the enhanced trophotrophic reactivity in a trophotropically tuned subject. Contrariwise, harassing the subject, a procedure known to induce a state of ergotropic tuning, may delay or abolish the immersion reflex, thus exemplifying reciprocal inhibition of the trophotropic system in this state.

Other aspects of Gellhorn's theory are discussed subsequently, relevant to the central issue of the study in this chapter. (Gellhorn, 1968).

Sudden fear in man may evoke bradycardia which may be followed by fainting (Engel, 1950) or even sudden death (Wolf, 1964).

The excellent review of Fara and Catlett's (1971) study of cardiac responses and social behavior in guinea pigs provides a basis for exploring further behavior-related cardiac phenomena in mammals. Adams et al. (1968) have recorded marked bradycardia in cats prior to responding to attack by a conspecific.

There may, however, be enormous species and individual differences. A normal dog, socialized to people, will show a marked bradycardia when handled by a person, in contrast to the tachycardia reported for the guinea pig by Fara and Catlett (1971). Does this imply that some species have more or less vagal or sympathetic tone than others? Also, what effects might domestication, early handling stress (Denenberg, 1967), socialization (Scott and Fuller, 1965), or the lack of socialization in the dog have on such cardiac responses.

Heart rate is, to date, the most accessible, if not the most accurate, measure of changes in autonomic activity—constituting a quantifiable physiological measure underlying emotional responses in various social contexts.

Lacey and Lacey (1970) present additional data on the bradycardia phenomenon in man, as well as some startling, if not controversial, conclusions. Transient bradycardia is regularly recorded during orientation and attention. Those tasks requiring only simple environmental reception produce significant cardiac deceleration, while difficult tasks, entailing internal cognitive elaboration of a problem-solving sort or requiring exposure to noxious stimuli, produce massive cardiac acceleration. An activation response (getting ready to respond to a conditioned signal), an anticipatory cardiac deceleration, also occurs, which is an indicator here of high motivation.

Israel (cited in Lacey and Lacey, 1970) has identified an intriguing relationship between this bradycardia of attention and the "cognitive style" of an individual. So-called sharpeners pay attention to everything, focus on differences rather than similarities, and habituate slowly (a description that would fit most wild canids). Other human subjects belonged to the category of levelers who tend to make global judgments, habituate rapidly, and are inattentive to many details of the environment (a description which partially fits the average domesticated dog). The levelers show a lesser attention bradycardia than the sharpeners. The importance of this study is not the analogy with canids, but rather, the demonstration that physiological lability parallels the lability of attitude manifest overtly toward the external environment. Central nervous system activity and behavior (including temperament and cognitive style), therefore, appear to be under far greater control of autonomic activity than was hitherto supposed by western psychophysiologists, although this fact has long been recognized though not quantified by Pavlovian physiologists (see Kurtsin's 1968 overview of corticovisceral and viscerocortical interrelationships).

Lacey and Lacey (1970) go on to show that this cardio-deceleratory response is correlated with Grey Walter's DC shift in the EEG activity—the contingent negative variation (CNV) which reflects the organism's readiness to respond. Apparently, the greater the CNV, the greater the cardiac deceleration.

The question now arises as to which comes first. Research by Bonvallet *et al.* and Zanchetti *et al.* (cited in Lacey and Lacey, 1970) shows conclusively that cardiovascular afferent input to the CNS results in a change in CNS activity. For example, elimination of glossopharyngeal and vagal input (to the bulbar inhibitory area) leads to poststimulation cortical activation. Thus viscerocortical afferents have an inhibitory effect on the CNS.

Lacey and Lacey (1970) conclude that

> The temporary hypertension and tachycardia observable in acute emotional states and in "aroused" behaviors of all sorts may not be the direct index of so-called "arousal" or "activation" they are so often considered to be. Instead they may be a sign of the attempt of the organism instrumentally to constrain, to limit, and to terminate the turmoil produced inside the body by appropriate stimulating circumstances. Moreover, to quote from an earlier statement of ours,
>
> ... if increases in blood pressure and heart rate signal a physiological attempt to restrain excitatory processes, then it seems likely that their diminution, absence, or conversion to blood pressure and heart rate decrease signify an absence of this restraining process and, therefore, a net increase in excitation: a lowering of threshold, a prolongation of the impact of stimuli, an increase in spontaneous activity, and the like.
>
> It is this last interpretation, and derivatives from it, that appealed to us as indicating the most strategic and dramatic approach to demonstrating the role of cardiovascular activity in the behavior of intact humans, and to challenging current activation theory, which has long seemed to us to be a grossly oversimplified view of the role of autonomic and skeletal activity in behavior.

(They cite research by Bonvallet *et al.* showing that distension of the carotid sinuses could cause a dog or cat electrocorticogram to appear as though the animal were sleeping.) Thus, the possibility arises that increases in blood pressure and heart rate may be physiological attempts to restrain excitatory processes rather than serve as a direct index of arousal or activation. Their logical deduction of the reverse is that decreased blood pressure and heart rate signify an absence of this restraining mechanism, a net increase in excitation, and a prolongation of the impact of stimuli.

Graham and Clifton (1966) have come, independently, to a similar conclusion *vis* that an increase in heart rate and an increase in blood pressure lead to inhibition of cortical activity and would presumably be associated with reduction in sensitivity to stimulation (as in defensive–protective reaction). Conversely, they theorized, heart rate decreases should be associated with increased sensitivity to stimulation (as in orientation or taking in the environment).

Lacey and Lacey (1970) also observe that:

We are doing chronic animal studies now to try to demonstrate that this visceral afferent feedback pathway does indeed operate in the way we think it does, in the experimental situations we have described. Many problems have yet to be solved and much work done before really effective demonstrations can be made.

In the meantime, the cardiovascular system commends itself to psychiatric study, not as a nonspecific index of arousal or emotion, but as a highly specific and apparently quite delicate response mechanism, integrated at the highest levels with the affective and cognitive variations among people, and revealing specific personal idiosyncrasies in the way people deal with their external world. (Lacey and Lacey, 1970, pp. 225–226.)

Socialization of the dog to man has a profound effect on heart rate. I (M. W. Fox) recall in veterinary practice being surprised at the extreme bradycardia and frequent arrhythmias of hyperactive little dogs presented for clinical examination; one would have anticipated tachycardia. It was only later that I fully appreciated that my examining and auscultating the dog was the cause of such a dramatic cardiac phenomenon.

Pavlov (1928) observed that the presence of a human being could easily inhibit an ongoing conditioned response, so he had to isolate his animals for study. Later he found that specific responses were evoked by certain people, and he postulated the existence of a social reflex in the dog.

Research on the involvement of the autonomic nervous system in behavior and in the establishment of conditioned reflexes and experimental neuroses in dogs was made possible with the advent of electronic physiological monitoring in the early 1940s. Gantt (1944), a student of Pavlov, was the first to demonstrate that the autonomic nervous system was easily influenced by the presence or absence of human contact, even when no overt behavior was manifest. One psychotic dog with catatonic symptoms, had a normal resting heart rate of 200 beats/min, but in the presence of a human being it would fall to 12 beats/min. Cardiac arrest, sometimes for as long as 8 sec, was recorded in this dog when in the presence of a person (in spite of which it lived for 14 years in the laboratory) (Newton and Gantt, 1968).

Subsequent research by Gantt *et al.* (1966) demonstrated cardiac changes in the rabbit, cat, dog, guinea pig, opossum, and monkey caused by the presence of people. Two consistent observations were reported in the dog: an increase of 10–80 beats/min when a person entered the room where the animal was and a decrease of 5–40 beats/min when petted.

Royer and Gantt (1961) subsequently showed that the cardiac response to petting in the dog was not an automatic reflex reaction. Several dogs would only show petting bradycardia with a familiar handler; a stranger evoked no response or sometimes tachycardia. Familiarity with the person is therefore of major importance. Scott and Fuller (1965) noted that there were differences among breeds in response to a person entering the experimental room, bradycardia being seen in cocker spaniels and tachycardia especially in basenjis. The former breed, especially selected for passivity and crouching, may be demonstrating a response comparable to fear bradycardia (see discussion later).

In a very original series of experiments, Lynch (1970) demonstrated that the bradycardia evoked by petting in dogs was not a simple reflex but was rather an integral component of an emotional reaction toward the person. He showed that tactile contact has a potent effect on the cardiovascular system of dogs to the extent that fear or pain (shock-induced) responses may be blocked, and the usual tachycardia to such conditioned stimuli was inhibited.

It has long been known that "contact comfort" of a hurt child (or pet animal) by the parent (or owner) "makes the pain go away." Lynch (1970) presents physiological evidence in support of this objectively unlikely phenomenon. The quieting effect of the presence of handler or familiar keeper in many species of domestic and wild captive animals is explained by these findings as is perhaps the potency of reward to a dog in a mere stroke from its master.

Unfortunately, neither Gantt nor Lynch considered the paradox of bradycardia evoked by contact in dogs that were afraid of people or unsocialized (feral) and the bradycardia they demonstrated in socialized dogs that were friendly toward people. A twitch on a horse or muzzle on a dog may evoke a comparable calming effect as does the presence of a handler in more tractable animals. The physiology of restraint and potentially harmful consequences of excessive autonomic reaction to enforced restraint or tonic immobility induced by pain or fear warrant further study. Furthermore, genetic influences (selection for domestication) in addition to such early experience variables (socialization to man) have not been studied, hence the emphasis in this study on wild (undomesticated) canids and unsocialized (feral) coyote × dog hybrids.

Contact comfort evokes a clear parasympathetic arousal (as reflected in bradycardia). A dog will work in order to get such a reward from its handler; petting a dog conditioned to respond to pain or react fearfully to a conditioned signal will inhibit sympathetic arousal and tachycardia. What role might this phenomenon have in development and in establishing social bonds between man and animal?

Lynch (1970) reports preliminary evidence that the cardiac response to petting develops somewhere between 12 and 16 weeks of age in the dog, sometime after the critical period for socialization (which is from 4–12 weeks of age). He found that the presence of a person (without contact) caused a 40% reduction in heart rate in a distressed and socially isolated puppy. Similarly, Liddell (1954) found that if a kid or lamb was exposed to a stressful environment in the presence of its mother, it could readily resist the stress; sibling twins stressed in the absence of the mother succumbed rapidly, and some showed persistent behavioral disturbances into maturity.

Thus, the question now arises concerning the lack of tactual contact on behavioral development, on the establishment of social bonds, and on subsequent resistance to stress.

Contact (petting or maternal care) would seem to facilitate the attachment or socialization process. Deprivation may cause severe and long-lasting behavioral and physiological changes in man and other primates (Harlow, 1959; Bowlby, 1953; Spitz, 1950). Tactile contact appears to be physiologically and psychologically beneficial (if not essential) for normal development. A dependence upon parasympathetic stimulation (evoked by the mother or handler) may also facilitate socialization. In the absence of such other-directed stimulation, isolated animals (puppies and infant rhesus and human infants) will effect self-stimulation (tail- or thumb-sucking, self-clutching, and rocking). The need for such stimulation in immature organisms may be important for the maintenance of normal physiological homeostasis. Without it, development is impaired. Also, if deprived of contact, stressful stimuli may not be adaptively countered by a parasympathetic reaction (as demonstrated by petting bradycardia [Lynch, 1970]). Stress factors (such as hospitalization) may then be intensified. Therefore, this is of considerable clinical significance in veterinary medicine and human pediatrics. It is also relevant to the husbandry of farm animals where stresses associated with transportation, weaning, and maternal separation are major causes of livestock losses and impaired growth.

Experiment I.
Development of Heart Rate

The heart rates for four breeds of dog (beagle, wirehaired fox terrier, Shetland sheep dog, and basenji) were recorded at weekly intervals from 1 to 5 weeks of age. Data from ten subjects of each breed were then averaged to give an overall view of rate changes with increasing age. Individual and breed differences were not analyzed (see later), since the purpose was simply to collect data representative of a heterogenous, pooled sample.

Subjects were held in the experimenter's arms and the heart rate measured using a stopwatch and stethoscope.

From an average rate of 220 beats/min at 1 week of age, there was a rapid decline to 192 beats/min by 3 weeks. After this age, the resting heart rate gradually increased to an average rate of 208 beats/min at 5 weeks. (See Figure 1.)

Suspecting that this developmental pattern was due to a gradual increase in vagal inhibition, data were collected from several mongrel and beagle dogs from birth to maturity (1–2 years of age), following injection (IM) of atropine sulphate (0.5 mg/kg). This

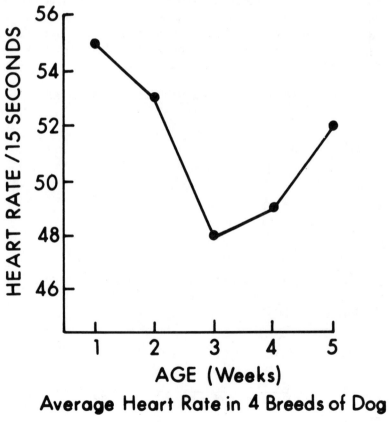

Average Heart Rate in 4 Breeds of Dog

Figure 1. *Average heart rate in four breeds of dog.*

would, by selectively blocking the vagal effect on cardiac activity, reveal at what age vagal inhibition begins to develop.

Interestingly, and as predicted, there was no significant increase in heart rate until after 10 days of age. The greatest release from inhibition occurred between 2 and 3 weeks of age, which tends to support the contention that the initial postnatal decrease in heart rate is related to the development of vagal inhibition.

The more gradual increase in heart rate following atropinization from 1 month of age through to maturity suggests that in addition to the relatively abrupt onset of vagal modulation of cardiac activity, there is also a more gradual increase in vagal tone. (See Figure 2a.) This latter phenomenon is to be anticipated in a species such as the dog, which is characterized in maturity by a high vagal tone compared to other species such as the rabbit, which have a greater sympathetic tone. Additional cardiac phenomena in young canids are shown in Figures 2b and c.

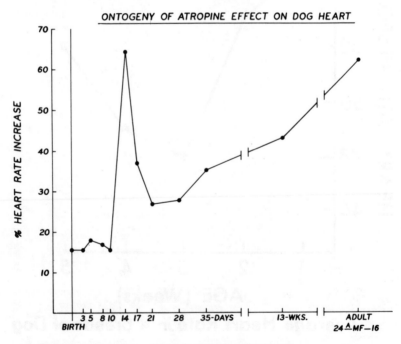

Figure 2a. *Ontogeny of atropine effect on dog heart.*

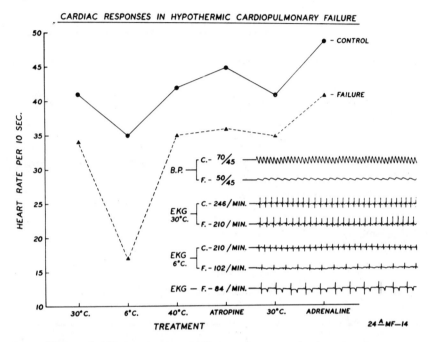

Figure 2b. *Various treatment effects on heart rate in normal and fading 2-day-old puppies; pronounced bradycardia in the latter. (From Fox, 1966.)*

FURTHER DEVELOPMENTAL OBSERVATIONS

Using direct auscultation and biotelemetric monitoring, heart rate development was recorded from birth until 8 weeks of age in five beagles, six coyotes, and three F_3 coyote × beagle hybrids. In all subjects, the previously described rate decrease between 2 and 3 weeks of age was confirmed as a developmental phenomenon independent of handling (petting or human contact).

Petting per se did not evoke a significant bradycardia until around 5–6 weeks of age; this effect was transient (3–10 sec) in contrast to the more sustained bradycardia recorded in adults.

Subjects with high resting heart rates as early as 1–3 days postnatally could be identified later as being the most active and

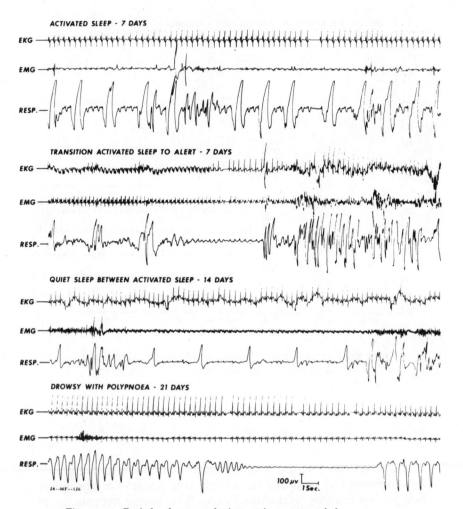

Figure 2c. *Periods of apnoea during various states of sleep are often accompanied by changes in heart rate.*

outgoing canids. Others with lower heart rates tended to be more passive or timid later in life.

This latter observation compares well with Scott and Fuller's (1965) developmental study of breed differences in heart rate. They found that the most outgoing breeds, wirehaired fox terriers and basenjis, had overall higher heart rates than the more passive shelties, beagles, and cocker spaniels. While the fox terriers were consistently different from 2 weeks of age, the other breeds did not individuate until after 7–8 weeks of age.

Lockwood (personal communication), working in my laboratory, concluded that the initial contact bradycardia of short duration in very young canids may be analogous to an orienting response (see also Lacey and Lacey, 1970). He also recorded a transient bradycardia when contact with a neonatal canid was broken; this "off response" may be an important autonomic indicator related to behavioral regulation of homeostasis, i.e., contact with mother for thermoregulation and nurturance must be maintained.

Confirming the findings of Scott and Fuller (1965), it was found that heart rate and sinus arrhythmia are highly interdependent in older canids. A slow heart rate (i.e., high vagal tone) is arrhythmic, and conversely, a high rate is regular without any arrhythmic inclusions. Scott and Fuller (1965) demonstrated breed (genetic) differences in these interdependent physiological indicators of autonomic activity and emotionality. It will be demonstrated shortly how early experience (handling stress) may also influence cardiac activity in later life.

Experiment II.   Can Early Experience Influence Cardiac Activity?

Research on early handling stress in rodents (Denenberg, 1967; Levine and Mullins, 1966) has shown that the adrenal–pituitary axis can be affected so that reactions to physical or emotional stress are more graded and adaptive than in control animals. The latter, unstressed in infancy, tend to overreact to stress. Early handling in rodents also enhances learning ability under stressful conditions, since emotional arousal is less than in controls and so does not interfere with performance. Resistance to cold exposure, starvation, implanted neoplasms, and viral leukemia has also been shown to be greater in rodents subjected to optimal handling stress in early life. Even more remarkable is that this phenotypic modification is transmitted nongenetically over several generations (Denenberg and Rosenberg, 1967) and may be instigated prenatally by handling the pregnant mother (Joffe, 1969).

The findings imply that environmental influences (i.e., stress) during a critical period in early life have profound and enduring consequences and may well influence sympathetic–parasympathetic tone as well as determine emotional reactivity or temperament in later life.

Blizard (1971) has discussed the significance and possible clinical implications of individual differences in autonomic responsiveness (cardiac activity) in laboratory rats. He found that rats given handling stress from 1–7 days of age exhibited diminished cardiac response to handling and to novel stimuli in adulthood compared to nonstressed controls exposed to these stressors for the first time as adults. Environmental influences early in life, therefore, have a clear effect on emotionality and, because of the reported influence on heart rate, may enhance trophotropic or parasympathetic tone in laboratory rats.

Theoretically, at least, the autonomic–neuroendocrine interrelationships and temperament per se may be similarly modified in the dog. The potentials for such phenotypic engineering may also have considerable practical value in raising dogs for stress resistance, an important consideration for military and peace-force use.

Given, then, that vagal tone is not already established at birth in the dog, it might be possible to modify sympathetic or parasympathetic tone by enhancing the former via handling in early life, as demonstrated in the aforementioned studies with laboratory rodents.

Therefore, it was decided to evaluate the physiological and behavioral effects of early handling in domesticated dogs. Four litters of random-bred pups were used, half of each litter being used as nonhandled controls (eight in all) and the remaining eight pups subjected to the following regime from birth until 5 weeks of age.

MATERIALS AND METHODS

This experiment was designed to determine the effects of differential rearing on several aspects of behavior and development of the dog. Sixteen dogs were studied (eight control, eight handled). The handling procedure and behavioral tests are described in Appendix I.

RESULTS

No significant difference in body weight gain was observed in the groups, nor were significant differences in organ weights or in total brain weight observed. When subjects were tested at 3 and 4 weeks of age by a variety of neurological responses (Fox, 1964), no significant differences were observed. Some handled pups, however, showed slightly superior coordination while standing and walking at 4 weeks. Histological examination of motor, occipital, and frontal cortex and vestibular neurons revealed inconsistent differences in neuronal size, but no differences in cell density were observed. Vestibular neurons in four handled subjects were significantly larger and contained more chromatin than those in the control or isolated pups; the total population of neurons observed in serial sections contained fewer small-sized neurons. The Meynert cells of the fifth layer in the occipital cortex and those of the fifth layer in the frontal and auditory cortices (large pyramidals) appeared larger in three handled pups compared with littermate controls, and they contained more Nissl substance. Heart rates were recorded in the control and handled groups at weekly intervals by auscultation and at 5 weeks of age with surface electrodes, at which time subjects were not handled (see Figure 3). The marked differences between the handled and control groups were apparent from the second week onward. Normally there is a decrease in heart rate from 2 weeks of age onward, owing in part to an increase in vagal tone (discussed earlier) which was seen in the control subjects; but in the handled group, cardioacceleration was seen from 2 weeks of age onward, and at 5 weeks the heart rates in all handled subjects were greater by 60 beats/min indicative of greater sympathetic tone.

The average percentage of epinephrine per adrenal gland was found to be 63% ± 7 in the controls. There was no change in the total amino content of the adrenals due to handling, but there was a significant increase in epinephrine. An increase in triglyceride content of the kidney and adrenals and a decrease in the liver were found in the handled pups compared with their controls, which

SLEEPING EEG OF HANDLED AND CONTROL 5-WEEK OLD PUPS

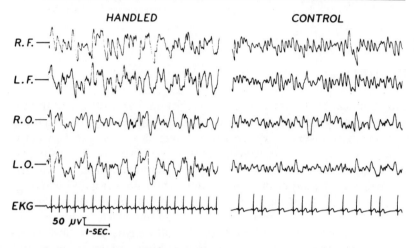

Figure 3. *Sleeping EEG of handled and control 5-week-old pups. Note greater maturity of EEG of handled pup (higher amplitude) and also faster resting heart rate.*

may indicate an increase or change in lipid metabolism as a result of early handling.

BEHAVIORAL OBSERVATIONS

Extreme differences were not found among individuals in the same group, and this surprising uniformity facilitated comparisons between the two differentially reared groups. These data have been summarized in Table I. Generally, the handled pups were hyperactive, highly exploratory, very sociable toward humans, and dominant in social situations (e.g., play) with their peers. The handled subjects showed the greatest distress vocalization immediately after the handler had entered the testing arena and removed the cloth and toy. In contrast, the control subjects were little distressed by this interference but showed great emotional arousal when first put into the arena. In the barrier test situation, the handled pups performed best in that they required fewer trials to negotiate the de-

Table I.
OPEN-FIELD BEHAVIOR IN DIFFERENTIALLY REARED DOGS[a]

Subjects (n = 8)	Specific stimulus		Random	Activities		Preference	Distress success rates	
	Cloth	Toy		Exploratory (nonspecific)	Distress (vocalization)		Right Closed	Left Closed
Control	78.0	37.2	32.5	152.3	27.2	R. 1.8	41.5	41.5
	76.0	7.4	68.0[b]	232.0	2.3	L. 2.3		
			34.0	182.6	10.0			
Handled	52.5	38.25	22.0	187.25	17.2	R. 1.6	93.0	96.5
	49.75	47.0	183.0	117.0	75.5	L. 2.2		
			13.0	190.25	8.5			

[a] Isolation for 4 to 5 weeks; handling from birth to 5 weeks.

[b] Figures in italics indicate observations when cloth and toy are removed from arena. All figures represent average time in seconds during three 5-min test periods, in which arena is full, then specific stimuli removed and replaced after 5 min.

tour, whereas the control pups reacted more slowly and showed distress vocalization in this situation (Table I). In the control group, therefore, emotional arousal in this situation prejudiced problem-solving ability.

Recordings of EEG of all the handled pups showed a greater amplitude during light sleep than those of the control subjects (Figure 4). As amplitude increases with age, it may be presumed to be an indication of greater maturity in the handled subjects. No difference in amplitude or fast frequency components were observed between the handled and the control pups in the alert state. A similar observation has been made on EEG maturation in kittens following early handling (Meier, 1961).

From these findings a pilot project was set up in cooperation with the Biosensor Research Project, U.S. Army Veterinary Corps. Several litters of German shepherd pups were given the early handling stress from birth to 4 weeks of age, each subject receiving only 1 min on a 45 rpm teeter-totter that gave both lateral and angular vestibular stimulation and 1 min cold exposure at 37°F.

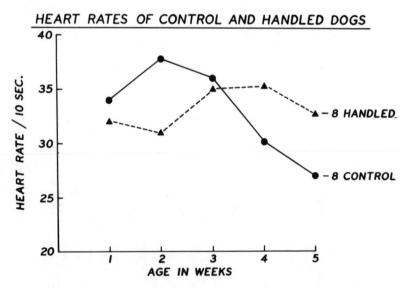

Figure 4. *Heart rates of control and handled dogs. Note sustained higher heart rates in handled pups.*

Preliminary results so far are extremely promising. As adults, such dogs have reportedly superior stamina under conditions of heat stress in the tropics. Data on disease resistance and longevity are not yet available. But a word of caution should be added. As Ginsburg (1968) has shown, working with different inbred strains of mice, the same handling stress may vary from one strain to another—having no demonstrable effect in some, enhancing others' later stress resistance, or lowering stress resistance in others. In a more heterogenous dog population, therefore, early handling may produce comparable results; thus, the quality and quantity of stress must be carefully regulated on an individual basis. Ideally, a suboptimal handling regime should be instigated (e.g., 1 min teeter-totter stimulation followed by 1 min gentle stroking, then 1 min cold exposure followed again by 1 min gentle stroking).

Under natural conditions, a wild canid mother may leave the cubs for extended periods, and they would be exposed normally to more stresses in early life than the relatively overswaddled home-raised litter.

Several breeders over the past 2 years have set up a simple handling stress regime using half of each litter, the other half being controls. Based upon subjective reports, all are enthusiastic and believe that this method of phenotype enhancement is a valid phenomenon (which controlled laboratory research has already demonstrated!) and a valuable addition to the usual rearing practices.

In fact, there is less originality in the application of this phenomenon in improving the temperament of domesticated animals. As Denenberg and Whimbey (1963) have suggested, prenatal handling or gentling of the pregnant mother and early postnatal handling of the offspring were probably practiced unwittingly as a part of animal husbandry early on in the domestication of both farm and pet animals.

The marked difference in the age-related changes in heart rate in the four breeds of dog described earlier and the control subjects in this latter study warrant scrutiny and explanation. The controls received regular daily contact with people and were well socialized and more used to being handled than the pups of four different breeds that were handled only at weekly intervals for recording growth and heart rates. The increase in heart rate in these dogs after 3 weeks of age (at the beginning of the critical period of

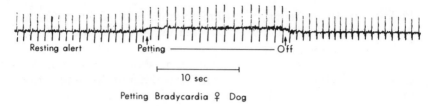

Resting alert　　　　Petting ————————— Off

|———————————|
10 sec

Petting Bradycardia ♀ Dog

Figure 5.　*Petting bradycardia in a female dog.*

socialization) may be a reaction to being restrained and handled. Such tachycardia, indicative of flight or emotional arousal (fear), has been recorded in several dogs of various ages. In socialized canids, however, a slowing of the heart rate occurs (see Figure 5). This contact bradycardia is discussed subsequently. It should be emphasized that the tachycardia that developed in the handled pups after 2 weeks of age was not correlated with fear or other emotional disturbance. On the contrary, these "super-socialized" puppies never resisted restraint or handling. Most significantly, they also manifested tachycardia even when resting unrestrained, with EKG being recorded by surface electrodes rather than by stethoscope (see Figure 4).

Clearly, such variables as socialization, habituation to handling, restraint and the method of measuring heart rate (either with surface electrodes to a polygraph or via stethoscope) are significant. Subsequently, therefore, heart rates were recorded with a bio-telemetry transmitter directly through a receiver onto a physiograph.

Experiment III.
Individual Differences in
Heart Rate and Temperament

The next question was to follow up the marked effects of early handling on cardiac activity and behavior (i.e., phenotype modification) with a study of individual differences within litters of canids given no handling stress. Individual differences in heart rate, if present, may correlate with other measures of emotionality and behavior and be an indicator of autonomic tuning or sympathetic tone independent of environmental (stress) influences; such individual differences, which can be created by handling stress, may also be genetically predetermined as part of a young animal's basic psychophysiological constitution or temperament.

If such individual differences could be demonstrated, how enduring might they be and how, under natural conditions, could they influence disease resistance, fertility, and other eventualities demanding adaptation to some form of stress?

Also, if there is consistency in such individual differences in heart rate and a high correlation of this measure of autonomic activity with other measures (of behavior, emotional reactivity, and learning ability), we may have an extremely sensitive index or test protocol for puppy evaluation. Such early evaluation has great applicability for screening litters of puppies and selecting individuals at an early age for special rearing and training programs (e.g., military dogs, guides for the blind, etc.).

This possibility has been explored from a somewhat different perspective in man. Thomas *et al.* (1970) have devised a battery of tests to evaluate the emotional/autonomic reactivity of human neonates. From these measures, they were able to make certain predictions as to what the later personality or temperament would be; these predictions were verified in this longitudinal study, the subjects now in their late teens. From a practical point of view, relative to dog and man, such early temperament predictions could enable one to instigate the best possible rearing conditions in order to optimize potentials and to control for individual traits that are maladaptive (such as overreactivity to novel stimuli, distractibility, emotional disturbance following change in routine or in the absence of immediate reward or comfort).

An initial pilot study was conducted on two captive adult wolves at the Naval Arctic Research Institute, Point Barrow, Alaska. Working in collaboration with Dr. G. E. Folk, biotelemeters were implanted intraabdominally into two yearling wolves, the highest- and lowest-ranking members of a captive pack being selected. Recordings were taken constantly for 48 hr to monitor possible circadian variables in activity. At all times, the heart rate and overall motor activity was greater in the highest-ranking wolf.

It was tentatively proposed (Fox, Folk and Folk, 1970) that such marked physiological differences might be predetermining factors for temperament and social rank.

With these promising findings, the following experiment was designed to explore further the possible significance of individual differences in heart rate correlated with other behavioral measures in unsocialized 6 to 8-week-old wolf cubs. This species was chosen in order to rule out any potential anomaly in vagal tone which, in a domesticated species, together with socialization effects, might have a significant influence on cardiac activity.

In the present investigation, a temperament or reactivity profile was made for each cub in two captive litters (as detailed by Fox, 1972), and the social rank of each cub was ascertained. Cubs of high, low, and intermediate scores were then selected for biotelemetric monitoring of heart rate under various test conditions; some of these subjects were then used in a stress study, the response being determined by changes in plasma cortisol levels over time.

MATERIALS AND METHODS

Twelve wolf cubs from two litters were studied between 6 and 8 weeks of age. A series of tests, described in detail in a recent study of individual differences in behavior of wolf cubs (Fox, 1972), were run (see Appendix II).

RESULTS

The scores for group and individual prey-killing, dominance (averaged scores from repeated tests), and the reactions to the novel stimulus are detailed in Table II. The close correlations between social rank, prey-killing, and exploratory behavior, as demonstrated earlier (Fox, 1972) are clearly evident in these two litters. In this earlier study, the question of social facilitation versus prior experience was posed since the group prey-killing test was conducted after the individual tests. The data in Table II show that social facilitation may indeed be present in the group test and

Table II.

SCORES FOR GROUP AND INDIVIDUAL PREY-KILLING, DOMINANCE, AND
REACTIONS TO NOVEL STIMULUS

	Group prey-killing	*Individual prey-killing*	*Dominance*	*Exploratory behavior*
Litter I				
ScBl ♂[a]	5	5	5	5
Bl ♀[a]	5	5	4	4
ScBr ♂	4	4	4	5
RcBl ♀[a]	4	4	3	5
UcBr ♀	4	5	3	5
CCBr ♂	4	3	3	4
CCBr ♀	4	4	3	5
KcBr ♂[a]	4	0	2	0
Litter II				
ScBl ♂[a]	2	3	5	0
RcBl ♂	2	1	3	0
BlUc ♀[a]	1	0	2	0
CCBl ♂[a]	0	0	1	0

[a]Selected for EKG and corticosteroid stress studies.

would account for the high scores of those cubs, which when
tested alone later, despite prior experience, had low scores. Differences in the magnitude of scores between the two litters are
marked. In the earlier study (Fox, 1972), the mother of Litter I
produced a similar outgoing litter the previous year, while the
lowest scoring litter in this same study was from the sister of the
dam to Litter II (both being sired by different males).

Changes in heart rate during the various handling procedures
are graphically represented in Figures 6 and 7. The most consistent
finding was the bradycardia associated with physical contact and

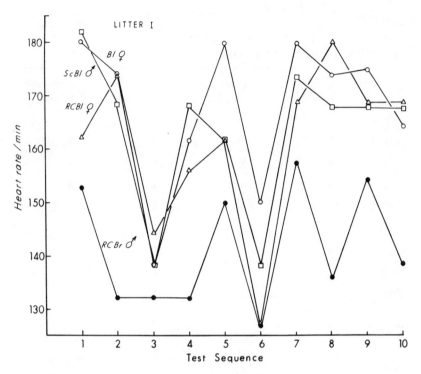

Figure 6. & 7. *Heart rates were recorded at the following ten stages in the test sequence. 1. baseline in arena; 2. observer enters arena; 3. & 4. contacts the cub; 5. observer backs off; 6. cub is picked up and held; 7. cub is left in arena and baseline rate recorded. 8. observer makes eye contact; 9. baseline rate recorded; 10. response to sudden noise.*

passive submission (or freezing). A second consistency was the higher baseline heart rates in the dominant cubs over the subordinates, a consistency which adds to the observations of Fox *et al.* (1970) who reported a higher heart rate over a 24-hr circadian periodicity study of a dominant yearling in contrast to its subordinate littermate. Bradycardia also occurred when the cubs were picked up. A less marked bradycardia was recorded when the experimenter remained passive in one corner of the arena and also when eye contact alone was made around the screen without the

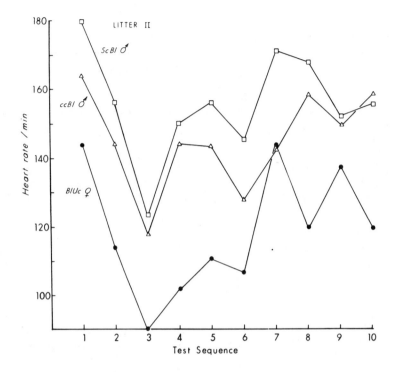

experimenter being in the arena. In both Litters I and II, the effect of
eye contact was most marked in the low-ranking cubs. Tachycardia, instead of the anticipated bradycardia, was recorded in RcBl ♀
of Litter I and CCBl ♂ of Litter II (see Figures 6 and 7), this reaction
being correlated with defensive–aggressive behavior.

Changes in μg% cortisol in the plasma of those cubs subjected
to the stress experiment revealed some signifcant trends (see Table
III). Generally, the highest-ranking cubs, which are quite capable
of adrenal response (according to ACTH treatment), delay the response for a time after the experiment has begun. In contrast, cubs
of intermediate social rank respond more quickly to stress in terms
of cortisol secretion and are capable of further secretion with ACTH
treatment. Subordinate animals are not very responsive in terms of
increased levels of plasma cortisol under stress, nor do they show
a capacity for adrenal secretory response to ACTH.

Table III.

SUMMARY OF WOLF ADRENAL RESPONSE TO STRESS

Animal	Condition	B.W. (lb/oz)	Cortisol in plasma (μg%)			
			5 min	10 min	30 min	60 min
Litter I						
α ScBl ♂	control[a]	15/6	22	70	61	36
β Bl ♀	control	15/3	39	69	190	65
ε RcBl ♀	control	15/12	95	100	88	50
ω RcBr ♂	control	16/8	38	25	24	26
Litter II						
α ScBl ♂	control	14/14	15	38	25	48
ε BlUc ♀	control	12/0	123	26	24	18
ω CCBl ♂	control	13/13	22	20	20	36
Litter I						
α ScBl ♂	ACTH[b]	15/6	92	100	184	47
ω RcBr ♂	ACTH	16/8	23	24	26	41
Litter II						
α ScBl ♂	ACTH	14/14	35	50	20	30
ε BlUc ♀	ACTH	12/0	60	36	25	34

[a]Control, serial blood samples taken from time of confinement with no further treatment.
[b]ACTH, treated with 20 IU Acthar (IM) at time of confinement.

It should be emphasized that the cubs of Litter II were more timid than most of those of Litter I (see Table II); in the responses of the dominant cubs of the two litters, for example, the dominant cub of Litter II resembled more a lower-ranking cub of Litter I.

These findings are discussed subsequently and correlated with individual differences in both cardiac responses and behavior as demonstrated in the various tests described earlier.

DISCUSSION

The behavioral test scores for prey-killing ability and exploratory behavior show a close correlation with rank scores for social dominance, especially in the more outgoing Litter I. These findings support the earlier study of individual differences in wolf litters of Fox (1972). In this earlier study, the question of social facilitation of prey-killing arose and was not completely answered. The present study was in part designed to answer this question, by giving the test with live prey to the whole litter prior to testing each individual. Several cubs had lower scores when tested alone with live prey than they had when tested with conspecifics.

The long-term significance of these tests should also be emphasized; cubs tested at 8 weeks could be identified on the basis of their scores when retested 1 year later (Fox, 1972). This serves to highlight the notion that the temperament of each individual essentially is determined innately. Or, at least the behavioral phenotype is formed by 8 weeks, so that by this age early experience has established a stable state (i.e., genotype–environment interactions are stabilized by 8 weeks). This basic temperament, which may be a constellation of autonomic tuning, adrenosympathetic response threshold, and emotional reactivity (see later), influences the way in which each animal responds to conspecifics, to prey, and to novel objects. Such experiences, in interaction with a particular temperament constellation during the first few critical weeks of life, determine subsequent social rank and behavior at least up to the first year of life. Certain allegiances within the pack may confound such predictions where a second ranking cub may become a subordinate outsider with maturity (Fox, 1972).

A consistent finding in the heart rate study was a higher resting heart rate, indicative of a greater sympathetic tone, in high-ranking wolves, while lower resting rates were characteristic of the most subordinate cubs. These findings support those of Fox *et al.* (1970) who found a higher heart rate and activity (in a circadian analysis of cardiac activity) in dominant than in subordinate yearling wolves.

The marked bradycardia manifest during parts of the test sequence with a handler is associated with freezing or passive immobility and has been recorded in many vertebrates (Belkin, 1968), as well as in unsocialized F_1 coyote × beagle hybrids during handling (see later). This bradycardia is to be distinguished from petting bradycardia in socialized dogs (Lynch, 1967) and other canids (see later). The contact bradycardia is of much longer duration than the orienting and attention bradycardia described in man (Lacey and Lacey, 1970) and is to be differentiated from the bradycardia which occurs after stress-induced tachycardia as immediately following a fight, for example (see later).

Candland *et al.* (1970) have studied changes in heart rates in squirrel monkeys in various social contexts, notably during aggression and establishment of dominant–subordinate relationships. They found that heart rate was related to rank on the status hierarchy by a curvilinear function; middle-ranking animals showed the lowest heart rates during test sessions but not during baseline measures in the home cage. In chickens (Candland, *et al.*, 1969), they found a similar U-shaped function; alpha and omega birds showed the highest heart rates during intraspecific test sessions, and middle-ranking ones had the lowest rates. These authors question why animals of both high and low rank should show higher heart rates than middle-ranking animals. It is not inconceivable that those of low and high rank have a greater sympathetic arousal associated respectively with threat or readiness to attack and flight or defense. Those of middle rank may have a greater parasympathetic tone related to passive submission or passive avoidance of conflict, so that escape-associated tachycardia is not manifest as in subordinates. (i.e. Tachycardia may be related to fight and flight in high- and low-ranking individuals.) Marked tachycardia was recorded in the squirrel monkeys during handling as distinct from the bradycardia of the wolf cubs.

Candland *et al.* (1970) found that when an individual experienced a change in social rank, there would be a predictable change in heart rate which correlated with its new rank position. These findings would be interesting to follow up and compare with wolves which might also show comparable physiological as well as behavioral changes as a consequence of a change in social relationships.

Murphree *et al.* (1967, 1969) reported lower resting heart rates in a timid strain of pointers, while a more stable strain had a higher resting heart rate. These findings compare well with the test scores of high- and low-ranking cubs which are analogous to the timid and stable temperament types that Murphree *et al.* studied. These authors were only able to demonstrate bradycardia during petting in the stable strain, a phenomenon associated with the reactions of socialized dogs to familiar handlers (Lynch, 1970; Lynch and Gantt, 1968). Bradycardia has also been reported in vertebrates in response to threat (Belkin, 1968). Again we have a paradoxical phenomenon; bradycardia evoked by both threat and petting but associated respectively with passive immobility or freezing and passive submission, both behaviors possibly involving parasympathetic relaxation and cardiac deceleration. The bradycardia associated with threat or stress may reflect the animal's readiness to respond; a vigilance state which has been correlated with changes in electroencephalographic activity (Lacey and Lacey, 1970).

Gellhorn (1968) has proposed that differences in temperament and behavioral reactions may be attributed to differences in sympathetic (trophotropic) and parasympathetic (ergotropic) tuning or tone. Schneirla's (1965) promising but as yet little tested hypothesis of biphasic sympathetic–parasympathetic processes in the ontogeny and organization of behavior is also relevant to this discussion. Both Gellhorn's and Schneirla's views could explain why animals react differently from a very early age to the same stimulus (e.g., live prey or novel object). Thus, the same stimulus would have different reinforcing properties between different individuals, and social stimuli (competition, threat, etc.) acting on cubs of different temperament could also have different reinforcing properties. Thomas *et al.* (1970), for example, have shown high correlations between the behavior of adolescents and their temperament constellations that were determined by a battery of tests during the first few months of life.

The Pavlovian concept of differences in nervous typology (synonymous with temperament) is also germane to this study. Typologies differ in terms of balance, flexibility (or dynamism), and relative strengths of internal inhibition and excitation. Thus, a strong balanced but dynamic typology or temperament would be characterized by adaptive responses to stress (neither overreactive nor hyporeactive) where the homeostatic balance between sympathetic and parasympathetic systems is well regulated (Kurtsin,

1968). Such an animal would show appropriate behavioral reactions to tests requiring passive inhibition, approach, or withdrawal.

The data on plasma cortisol secretion under stress and capacity to respond after ACTH treatment as well as the heart rate data accord well with the above concepts. It may be tentatively concluded at this level of analysis that dominant wolf cubs have a high sympathetic tone but do not overreact to stress (i.e., parasympathetic homeostatic regulation or balance is evident). Low-ranking cubs have a low sympathetic tone and show a hyporeactivity to stress. Those of intermediate rank have a higher sympathetic tone but tend to overreact to stress.

Follow-up studies over a 3-year period confirmed that these individual differences in temperament and rank order, established very early in life, were enduring provided the relationships with the litter were not disrupted (as by the death of one key individual or the introduction of a stranger).

These findings indicate that heart rate is a reliable index of temperament, but since both are determined genetically and experientially, changes in social relationships may affect heart rate. This was not determined in canids, but the work of Candland et al. (1970) with squirrel monkeys shows that a change in rank is correlated with a change in heart rate. They observed that

> Of special importance is the fact that when animals changed position on the (rank) order their heart rates changed accordingly, maintaining the curvilinear relationship. The implication of this result is that when dominance orders change, the heart rate of the animals changing position also changes. It suggests that rank determines the heart rate, rather than the reverse.

This latter conclusion is unwarranted however without a developmental study; their subjects were apparently unrelated and from diverse sources.

Candland et al. (1970) found that there is an inverted U- or J-shaped curvilinear function with the lowest resting heart rates in middle-ranking animals, in contrast to the wolf where rank correlates more directly with heart rate.

These authors ask, since they have also demonstrated a very similar correlation between rank and heart rate in chickens

Why would animals both high and low on the order show the higher heart rate? Or, conversely, why should midranking animals show the lowest heart rate? One possibility is that the noradrenalin/adrenalin ratio differs substantially in alpha and omega animals and has the result of generating different forms of behavior (flight or fight) while increasing the heart rate in both cases.

Since neither of these species are predators, the observations of Funkenstein (1955) might account for the differences in cardiac activity between wolves, squirrel monkeys, and chickens. He states that

> Experiments suggest that anger and fear may activate different areas in the hypothalamus, leading to production of nor-adrenalin in the first case and adrenalin in the second. Until more experiments are made, these possibilities must remain suppositions.

> Some of the most intriguing work in this field was recently reported by von Euler. He compared adrenal secretions found in a number of different animals. The research material was supplied by a friend who flew to Africa to obtain the adrenal medullae of wild animals. Interpeting his findings, J. Ruesch pointed out that aggressive animals such as the lion had a relatively high amount of nor-adrenalin, while in animals such as the rabbit, which depend for survival primarily on flight, adrenalin predominated. Domestic animals, and wild animals that live very social lives (e.g., the baboon), also have a high ratio of adrenalin to nor-adrenalin.

It may be, therefore, that canids having a higher amount of noradrenalin (associated with aggression and attack) would exhibit tachycardia in socially competitive contexts, and those with a greater sympathetic tone would be more aggressive or assertive and assume a higher rank than those with a lesser sympathetic tone (and lower resting heart rate). It should also be emphasized that passive inhibition rather than flight is the usual social reaction to threat in low-ranking wolves, and this might be correlated with a lower resting heart rate. This speculation was evaluated in a subsequent study (see later).

Contrary to the notion that high vagotonia is an attribute of athletes and of dogs with superior physical stamina, what adaptive value would high sympathetic tone have for the wolf? The higher sympathetic tone experimentally induced in domesticated dogs (described in Experiment II) showed that such animals were emotionally more stable, more exploratory, and assertive. In the social context of the pack, such an individual would be more likely to survive under conditions of food shortage in a severe winter, and such wolves are usually the only ones to breed in a pack. How individual variance or heterogeneity is maintained in litters of offspring to counter this unidirectional selection is an important question and deserves further study.

As Linn (1974) emphasizes, "the relative effect of ACTH and plasma corticosteroids on the psyche of the animal may well be determined by the state of the ergotropic–trophotropic balance of the hypothalamus (Gellhorn, 1968). Gellhorn (op. cit.) has theorized that ergotropic–trophotropic balance is closely interrelated with secretion of adrenal corticosteroids. Increased corticosteroids reduce sympathetic hypothalamic reactivity, cause a concurrent rise in parasympathetic hypothalamus reactivity and homeostatic balance is shifted to the trophotropic side."

Field observations show that the lowest-ranking wolves are the least likely to survive a severe winter (again reflecting the findings in nonhandled rodents being less resistant to terminal starvation or cold exposure than those handled subjects that have an induced high sympathetic tone). The data on plasma cortisol levels and response to ACTH imply that such canids are essentially incapable of mobilizing an adaptive stress-response. This contrasts the overreaction to stress in nonhandled rodents; it may also account for the low resting heart rates in low-ranking carnivores, and such prey species of comparable social rank (as studied by Candland et al., 1970) may have abnormally high resting heart rates.

Experiment IV. Heart Rate and Plasma Cortisol as Predictors of Temperament

The following is an abstract of an extensive study by my colleague and student, Dr. J. Linn, representing an attempt to employ the testing procedures used in controlled laboratory conditions to a more uncontrollable kennel-rearing facility operated by army personnel. (For details of test procedures see Fox, 1968, 1971; Linn, 1974; and *Appendix* III.)

The plasma cortisol level and heart rate of one hundred and nine twelve week old German Shepherd Dog puppies was monitored prior to, during and after undergoing a psychologically stressful experience. The physiological parameters were then examined for correlation with temperament of the dogs as they matured.

Because of large standard deviations there were no statistically significant correlations between either of the parameters measured and temperament. Plasma cortisol values showed a trend of prolonged elevation in response to stress in the puppies with the poorer temperament. (See Figure 8.)

There was a consistency in the results of the cortisol determinations. The mean plasma cortisol levels for baseline and fifteen minute post-stress showed a J shape with the highest rated pups and the lowest rated pups at the extremes of the J. Prior to testing the top rated puppies had the highest cortisol levels and fifteen minutes after testing the lowest rated puppies had the highest cortisol levels.

The mean heart rate of the group of puppies which had failing temperaments did not increase as much as the mean heart rate of the passing group when first put into the stressful situation. Their mean heart rate two minutes post-stress was higher than in the passing group. Thus, both parameters showed trends to prolonged physiological reaction to psychological stress in failing puppies. However, the range of one standard deviation on

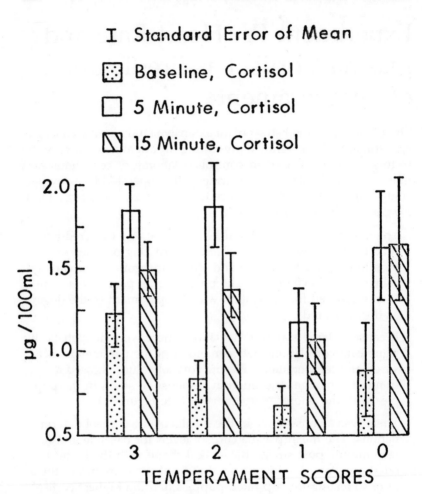

Figure 8. *(From Linn, 1974.)*

either side of the mean for passing/failing groups overlap too much to allow successful prediction of future temperament. (See Figure 9.)

A critical reevaluation of this potentially very promising screening procedure for early puppy selection reveals one important methodological omission. Resting heart rates were never

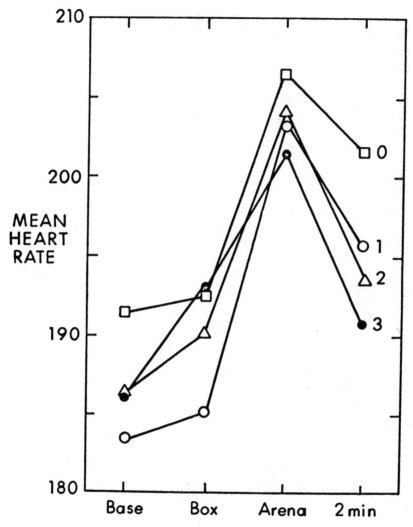

Figure 9. *(0–3 are temperament scores) (From Linn, 1974.)*

monitored biotelemetrically from puppies in their home cages. Base heart rates were recorded shortly after an initial jugular vein sample for plasma cortisol determinations had been taken (see Figure 9).

This prior stress served to segregate the pups on the basis of an inverted U function. The two groups of pups scoring best on temperament ratings (types 2 and 3) being intermediate between the reject pups (types 0 and 1) which showed relative bradycardia or tachycardia poststress. The greatest increase in heart rate occurred in the highest-ranking pups when placed in confinement (Box) prior to release into the novel environment (Arena). All four groups of pups showed a similar increase in heart rate in the arena.

The most significant finding is the heart rate recordings taken 2 min after exposure to the novel environment (i.e., recovery). Return to a resting level was graded inversely in relation to the temperament rating, the lowest rating pups having the highest heart rates, reflecting a sustained tachycardia following stress.

As in previous studies involving bradycardia in animals, insufficient distinction of behavioral signs was made between the bradycardia of fear and of being petted or restrained by the handler. Both friendly and fearful dogs may show passivity (passive submission); monitoring of muscle tone (EMG) may help differentiate these two states of passivity in future studies.

An additional factor which naturally reduced the variance between categories of test subjects was in the initial selection procedure; extremely timid pups were discarded initially and were never used in the study. Had they been included, a more natural, heterogenous sample could have made a significant difference in the final analysis of the data.

Experiment V. Bradycardia in Fear and Friendship

Bradycardia was evoked consistently in unsocialized (feral) canids (see Figure 10). A 20–30% decrease in heart rate from "resting-alert" was usually recorded. The normal baseline of resting-alert was taken, since general locomotor activity causes a considerable increase in heart rate.

Resting-alert heart rates were recorded biotelemetrically with the subjects alone in the testing cage and monitored via closed

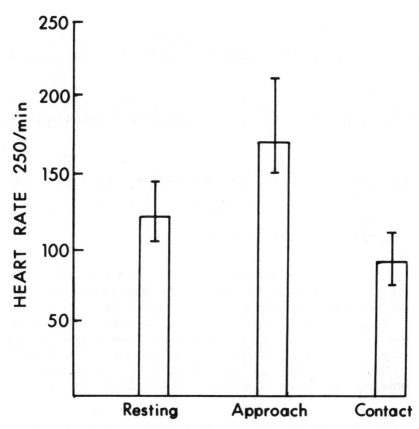

Figure 10. *Heart rate decrease with contact in unsocialized coyote × dog hybrids.*

circuit television. Heart rates were also recorded when a handler entered the cage and sat 6 ft away from the subject, remaining passive and avoiding eye contact. After 4 min, the handler stood up, slowly approached the animal and made physical contact (petting) for 1 min, and then slowly backed off and left the cage. Five repetitions for 3 consecutive days were made.

In the four unsocialized canids studied (F_1 generation coyote × dog hybrids), bradycardia developed in two of the subjects during repeated tests *preceding* contact. This may represent a conditioned autonomic response or expectation. (See Figure 11.)

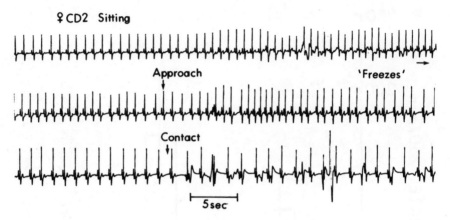

Figure 11. *Precontact bradycardia.*

Tachycardia was frequently recorded when the handler initially entered the cage and later when he stood up preparatory to approaching the animal. In the latter context, no subject attempted to escape although this cardiac response may reflect a readiness to escape—the flight response was blocked by the confines of the cage. In one subject, tachycardia was recorded as the handler withdrew after making contact; this was correlated with attack intention movements (presumably evoked by the flight of the han-

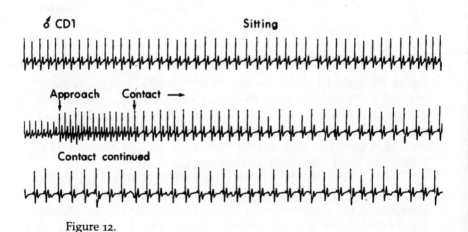

Figure 12.

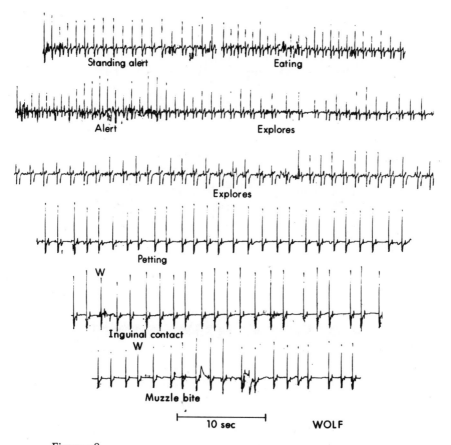

Figure 18.

or slight bradycardia (pleasure response?) were frequently recorded. When intimidated while eating, a sudden, transient bradycardia was occasionally recorded in association with freezing threat over the food and was more sustained following an actual fight (see Figure 19).

From these comparative observations, the sustained low heart rates during petting, assertion of dominance (eliciting passive submission), and contact in an unsocialized animal (producing freezing or tonic immobility) may be a common psychophysiologi-

Sitting

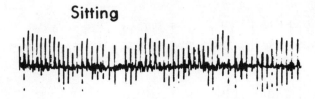

Threat

Attack

Recovery

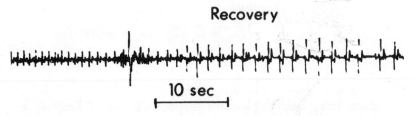

10 sec

Figure 19.

cal response. Some of these behavioral reactions in which sustained bradycardia have been recorded are illustrated in Figure 20. The possible evolutionary significance of this phenomenon is discussed subsequently.

Figure 20. *Various displays in which bradycardia may be recorded: (a) Tonic immobility in a frog (see McBride et al. (1969) for further details.); (b) tonic immobility in a wild canid (Dusicyon fulvipes); (c) passive submission in same subject after several months of handling or socialization; (d and e) bradycardia recorded in wolf cub in submission and in wolves, respectively, during more active submissive greeting; (f) when being pinned by muzzle by superior wolf; (g) passive submission in socialized jackal and maned wolf; (h) similarly demonstrate the behavior similarities between fear (tonic immobility and passivity) and passive submission while being handled—in both behavioral states, bradycardia normally occurs.*

General Discussion

These observations and the literature reviewed present a somewhat paradoxical picture of the relationships between cardiac activity and various behavioral states. Paradox is to be anticipated when one is dealing with a complex neurohumoral homeostatic system such as the autonomic nervous system. As Gellhorn (1968) ably demonstrates, paradoxical responses are often seen and can be explained on the basis of sympathetic–parasympathetic balance or ergotropic–trophotropic tuning in his terminology. Thus, an apparently paradoxical response such as bradycardia can occur in very different emotional states ranging from fear to pleasure.

An important factor in predetermining the type of autonomic response may be the degree of sympathetic and parasympathetic tone characteristic of the individual or species in question.

From the data presented, it may be hypothesized that wild canids (of high social rank) have a high sympathetic tone and a high parasympathetic tone. Low-ranking, more fearful animals generally have a lower sympathetic tone coupled with either a weak parasympathetic tone or an excess of vagal inhibition either of which may lead to maladaptive stress responses. In Pavlovian terminology, the latter two typologies would be characterized respectively as weak and imbalanced nervous typologies and the former as strong and balanced. His fourth category, strong but unbalanced, would be seen in those middle-ranking canids that showed less adaptive mobility in inhibiting excessive sympathetic or parasympathetic reactions but were able to adapt better to stress conditions than the weak and unbalanced types (Kurtsin, 1968).

The effects of early handling stress would seem to be creating a strong nervous typology with high sympathetic and parasympathetic tone.

Domestication, from comparative studies of wild and tame canids (and other species cited), may have selectively reduced sympathetic tone rather than enhanced parasympathetic tone. It would be naturally advantageous to reduce the fright-fight-flight responses of animals in the process of domestication [and as demonstrated by Beylaev and Trut (1975) the adrenal-hypothalamus axis responds in quantitatively different ways to stress in domesticated subjects].

Great individual differences are evident in dogs of different breeds and temperaments, however. These differences do not contradict the above hypothesis on the effects of domestication but rather point to the original question of ergotropic–trophotropic balance and dynamics or mobility in the Pavlovian sense. Such individual differences may still be manifest in the presence of reduced sympathetic tone or lowered ergotropic tuning.

In his two classic papers on the effects of domestication and selection on the behavior of the Norway rat, Richter (1954) states that

> Electrocardiographic records taken while the rats are being restrained show, in the case of wild rats, marked slowing of the heart—a definite bradycardia, and little or no change in the domesticated rats. Thus it appears that, as compared with wild rats, domesticated rats are much less vagotonic.

Is this conclusion warranted without studies of normal resting heart rates in wild and domesticated rats? If the flight response (and associated tachycardia) is not mobilized because of enforced restraint (analogous to passive inhibition in canids), then bradycardia might be a highly adaptive response.

Richter goes on to note that

> Domesticated and wild rats were subjected, by Dr. James Woods, to the same form of stress; either exposure to cold, fighting, or exhaustion from swimming. In the wild rat none of these forms of stress had any detectable effect on the amount of ascorbic-acid content in the adrenals, whereas in the domesticated rat it either reduced it to a very low level or else eliminated it altogether. In wild rats, the ascorbic-acid content could be depressed only by large doses of ACTH. These results would thus indicate that the wild rat is better able to withstand various forms of stress—because of the more active adrenals.

The wild rats clearly show a resistance to stress not unlike the superior stress resistance in rats given handling stress in early life. It would seem that we cannot, without further research, make a simple conclusion vis a vis vagotonia and effects of domestication. It may be highly adaptive for a wild animal to freeze or show passive inhibition (tonic immobility) in order to avoid either attack by a predator or by a superior conspecific. Thus wild rats and wild canids show bradycardia under forced restraint, and both wild and socialized canids show bradycardia when being petted in association with passive submission and friendly greeting. The latter behavior may have evolved, in the ethological sense (Chance, 1962), as a means of inhibiting or cutting off a potential attack by a social superior. This could account for the bradycardia occurring in two apparently unrelated contexts (see Figure 20).

Increased parasympathetic activity (salivation, slowing of heart rate, and increased peristalsis and secretion of digestive juices) has long been recognized as a response in infant animals and human neonates alike to maternal attention (Spitz,1949). This parasympathetic (pleasure principle) response in adult animals socialized to man is readily evoked by human contact and petting, as it is by one adult animal grooming a conspecific.

Absence of such stimulation in infancy may lead to impaired growth (marasmus) and increased susceptibility to disease in both human infants (Spitz, 1949) and dogs (Bleicher, personal communication).

Thus, parasympathetic stimulation, an important aspect of maternal care and neonatal physiology, may be linked with socialization and potentiate the development of emotional attachments. This has been elaborated upon in detail by Schneirla (1965).

In summary, therefore, contact with an animal socialized to man evokes relaxation and bradycardia—the classic psychophysiological response to petting or contact comfort. Contact with an unsocialized animal, however, evokes bradycardia associated instead with tension, freezing, or tonic immobility (see Figure 20). This latter cardiac response is probably an adaptive homeostatic mechanism to control for sympathetic hyperarousal. A comparable compensatory inhibition was recorded in a canid immediately after a fight with a conspecific (see Figure 19). Richter's (1954) conclusion that domestication reduces vagotonia therefore may be valid. High vagotonia would be an adaptive response to sympathetic overarousal which would occur in association with a wild animal's initial attempts to escape from restraint. Data on unsocialized canids support this view; marked tachycardia when approached but with escape blocked, bradycardia develops and is enhanced by physical contact (i.e., transition from freezing to passive submission). With a lower sympathetic tone in a tame or domesticated animal, restraint or human contact would not mobilize a compensatory homeostatic vagal inhibition.

But based on Richter's rather superficial data, vagotonia appears to be more a function of context than of domestication per se, since human contact evokes bradycardia in a socialized dog just as enforced restraint evokes bradycardia in an unsocialized trapped wolf (M W. Fox personal observation).

Domestication has more likely influenced the contexts in which vagotonic reactions occur and has lowered sympathetic tone and modified the adrenal–hypothalamic stress reaction threshold (a point demonstrated by Beylaev and Trut in foxes selected for docility over 14 generations). Consequently, with a lower sympathetic tone relative to wild species, the compensatory "phasic" vagal inhibition (rather than "tonic" vagontonia per se) would be less in a domesticated animal. Such a rephrasing of Richter's conclusions helps clarify the interrelationships between the adrenal-hypothalamic sympathetic/parasympathetic feedback system and would account for the sudden death syndrome (Cannon, 1956) in wild animals under restraint and in domesticated animals which have an abnormally high sympathetic tone which may trigger an excessive phasic vagotonia leading to a vasovagal attack.

This latter phenomenon, analogous to a Pavlovian "collision" of excitation and inhibition (but leading to a collapse of the organism rather than to a neurotic reaction), may account for the apparent absence of a plasma cortisol response to the stress of restraint in low-ranking wolves.

Kurtsin (1968) has reviewed extensive studies of temperament or nervous typologies of three different breeds of dog using Pavlovian conditioning procedures. He demonstrated clearly how different temperament types react to stress, either emotional or physical (e.g., total body irradiation). Susceptibility to stress and disease is greater in those breeds that essentially overreact, both autonomically and behaviorally, and lack adaptive behavioral and physiological inhibitory mechanisms. Future biomedical research might be advantageously focused on the role of the autonomic nervous system and sympathetic/parasympathetic interrelationships in terms of disease susceptibility and stress resistance in dogs of different breed and temperament. Again, as emphasized earlier, handling stress in early life might be instigated to improve the stress tolerance and disease resistance of susceptible breeds or individuals. As Kurtsin (1968) emphasizes, sympathetic overarousal to stress, mobilizing high levels of adrenocortical steroids, may interfere with immune responses and impair normal healing by reducing the phagocytic activity of leukocytes. Hyporesponsive individuals are no less susceptible either. Autonomic and emotional (temperament) variables must be considered in understanding individual susceptibility or resistance to disease (see also Astrup, 1968). Given the wide variety of dog breeds, this could be a fruitful area for future research.

Comparative studies of wild and domesticated cats (Leyhausen, 1973) and canids (Fox, 1971b) have shown that wild species are more active and responsive to novel stimuli, either visual, auditory, or olfactory. Domestication, therefore, may have affected the reticular arousal system and the response threshold of exteroceptors, a process which may have been mediated quite simply by lowering sympathetic tone. Gellhorn (1968) refers to this as ergotropic tuning and makes the following observation, which tends to support the above theory of the influence of domestication (artificial selection for temperament-emotional reactivity) on autonomic balance. A shift toward parasympathetic (trophotropic) predominance is proposed.

> Elimination of important afferent impulses tends to cause a shift in trophotropic–ergotropic balance to the trophotropic (parasympathetic) side. Thus, animals (cats) deprived of the organs of smell, vision and hearing show an increased reactivity of the trophotropic system indicated by a lesser general activity and an increase in the duration of sleep. Such animals show also behavioral reversal. Nociceptive stimuli, which in the normal cat induce ergotropic effects, exert trophotropic actions in cats which are blind, deaf and anosmic.

The drastically reduced sensory input is analogous to the raised response threshold of exteroceptors proposed above. This view differs from that of Richter (1954) who may have misinterpreted the bradycardia of wild rodents as a general (tonic) indicator of vagotonia, whereas in fact it was an acute (phasic) trophotropic reaction to stress. Its absence in domesticated rats could simply be because they were essentially more docile and were protected from such stress by having a lowered sympathetic tone and a greater trophotropic tuning (or tonic rather than phasic vagotonia).

Murphree *et al.* (1967) have reported changes in heart rate in their two lines of stable and unstable (shy) pointers under different conditions. As in the findings with wolves and German shepherds reported earlier in this book, the more outgoing stable line had a higher resting heart rate (approximately 120/min versus 80/min for the timid line). Contact with handler evoked a clear bradycardia only in the stable line, the decrease in heart rate being slightly greater when petted by a stranger. This latter result, not discussed by the authors, supports the hypothesis presented earlier that the petting bradycardia may be related to passive (friendly) submission, which in a socialized dog, may be greater toward a strange person.

Most interesting, however, was the lack of bradycardia in the nervous line of pointers. There was essentially no change in heart rate when the animals were being petted—neither with an unfamiliar nor a familiar person.

All subjects were restrained on a stand in order to monitor heart rates with a cardiotachometer. It is possible that this variable may invalidate any comparisons between these dogs and the freely moving unsocialized canids reported in this chapter. However, judging from the abnormal behaviors described in other contexts in the unstable line that they studied and the normal behavior of our subjects when human beings were not around, we are probably dealing with two very different behavioral complexes. The subjects used in the present study were behaviorally normal except for their fear of man (because they had never been socialized to man early in life). Consequently, human presence evoked flight and tachycardia, confinement triggered freezing, and some bradycardia, and physical contact resulted in marked bradycardia—normal physiological reactions. The unstable pointers from a genetically selected line had received much human contact in early life. Their sustained low heart rates, absence of tachycardia when approached and of any further bradycardia when contacted, and catatonia suggest an excessive vagotonia or inadequate trophotropic balance. Further studies have disclosed an increased incidence of atrioventricular heart block, which together with low resting heart rates, would suggest that these animals have an abnormally high vagal tone (Newton *et al.*, 1970). This conclusion is further supported by the recent finding (Lucas *et al.*, 1974) that these excessively fearful pointers lack the normal hippocampal theta rhythm (and had a tendency to sleep more than normal dogs). (This accords well with Gellhorn's conclusions cited earlier.) The inability of these dogs to habituate normally to novel stimuli was correlated with an absence of normal hippocampal theta activity during wakefulness. It is normally depressed (desynchronized) in the initial phase of an orienting response but returns to theta rhythm as habituation takes place. These dogs also manifested catatonia (inhibition of motor function), which can be evoked in normal dogs by electrical stimulation of the hippocampus.

A logical expectation would be that ergotropic rather than trophotropic enhancement would occur during such orientation/ attention processing. Novel stimuli, however, usually evoke bradycardia (as mentioned earlier in Fara and Catlett's study of the guinea pig). Lacey (1967) has described a similar attention bradycardia in man as a transient cardiac deceleration interpreted as a "preparatory" autonomic response preparatory to subsequent performance.

Graham and Clifton (1966) have also reported transient bradycardia in man as a component of the orienting response to novel stimuli. Orbrist (1968) has shown that under these conditions cardiac deceleration is accompanied by a decrease in respiratory frequency and amplitude. These orienting- and attention-related bradycardias thus differ from the fear bradycardia, which is of much longer duration and is usually accompanied by polypnoea (Hofer, 1970). Thus, the entrainment or influence of respiration on heart rate is probably not significant. In petting bradycardia in socialized dogs, muscular relaxation and decreased respiration are associated with a decrease in heart rate. In a fearful unsocialized canid, respiration may be increased but bradycardia is elicited following contact, and the animal is in a catatonic state of motor inhibition (see details later).

Conclusions: The Phylogeny of Submission— Bradycardia and Affection

The following theory is an attempt to resolve the apparent paradox between the bradycardia of fear and friendliness. In fact, there may be no paradox. The bradycardia associated with passive (friendly) submission may be a recent evolutionary development in the more social vertebrates from the bradycardia of fear, associated with passive inhibition, freezing, catatonia, or tonic immobility characteristic of more primitive vertebrates and invertebrates alike.

Consider the modes of reaction to threat or impending danger. Flight or attack with associated tachycardia or tonic immobility (playing dead), in many prey-species, is a highly effective antipredator adaptation. In such a behavioral state, the bradycardia is sustained (as distinct from the transient bradycardias associated with the orienting response and attention).

It is often stated that ontogeny recapitulates phylogeny, and in the context of the present theory, this may in fact be true. Tonic immobility is present in young mammals of many different species (deer, fox) and tends to disappear with maturity, whereas in more primitive vertebrates (rodents, reptiles), it persists throughout life. In the former species, it occasionally may be triggered by a sudden stress, i.e., the shock of hysteria evoking tonic immobility in an adult dog (see Figure 21) or human being.

This phenomenon has been related to animal hypnosis (Völgyesi, 1966) and to sleeplike states, which may bring about a restoration of homeostasis by de-arousal (as in conflict situations where displacement behaviors resemble certain actions preparatory to going to sleep [Delius, 1967]). It should be recalled that these sleeplike behaviors, associated with bradycardia, can be evoked by electrical stimulation of the carotid sinus (Lacey and Lacey, 1970).

In the evolution (phylogeny) of animal displays, it is well recognized that some that are present in one context may occur in other species in a very different context. For example, the defensive fear-grimace of primates and canids is thought to be the evolu-

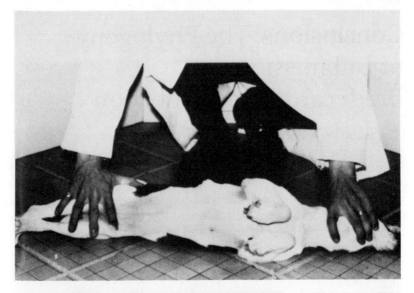

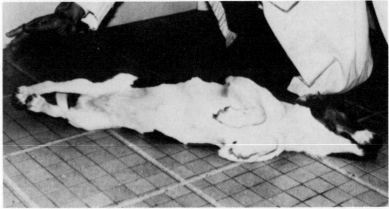

Figure 21. *Tonic immobility in the beagle from an extremely*
timid line.

tionary precursor of the submissive grin displayed in other species
as a greeting signal (Fox, 1971). A similar evolutionary emancipa-
tion may have taken place in the bradycardia of fear and contact-
affection.

If the bradycardia causes a change in behavior following some
intense noxious or fearful stimulus, i.e., decreased activity and

balancing of sympathetic overarousal, natural selection would favor inactivity if it was an effective antipredator strategy.

Similarly, natural selection in social contexts, as where a superior (alpha wolf or dog's master) evokes passivity and submissive displays, could operate on these same primitive psychophysiological (cardiovascular-CNS) processes. Chance (1962), for example, has emphasized how passivity may cut off a possible attack by a social superior in primates.

But this is not the end of these tentative correlations. A passive or actively submissive (greeting) dog or wolf will usually show several infantile behaviors, such socio-infantile actions being characteristic of more social, gregarious mammals (Fox, 1975). Such regression to an infantile mode forces us to consider the role of parental care in infancy. In socially dependent adults, there may be a persistence into maturity of some aspects of the psychophysiological dependency of infancy (discussed earlier, where parental behaviors evoke parasympathetic arousal and and sympathetic de-arousal in accordance with Schneirla's [1965] biphasic theory of behavioral organization).

This initial dependency of the infant on parasympathetic stimulation (contact comfort) leads to a canalization of development toward social dependency in later life and a persistence of analogous physiological changes in contexts of submission, greeting, and care-solicitng. In other words, the essentially inner-directed homeostatic function of parasympathetic arousal and sympathetic de-arousal (as in the bradycardia of tonic immobility) of primitive vertebrates becomes other-directed in higher vertebrates and instead occurs in social contexts. *This psychophysiological dependency upon others forms the basis for the development of affectional bonds.* Petting a tame snake or crocodile is a social contingency without reward, whereas a a dog will work for such reinforcement.

The evolution of this psychophysiological dependency is verified by the wealth of data on experimental deprivation of contact-reinforcement in infant mammals. They will not thrive and may even die. Bereavement and depression following the loss of a mate or of a close companion in adult mammals of many species also demonstrate how pervasive this psychophysiological mechanism for social dependency is. Without such other-directed dependency, the adult organism is relatively asocial and leads a more or less solitary life. Deprivation in more sociable species (dog, rhesus monkey, and man) will impair the normal development of dependency relationships. Such overly inner-directed individuals have an extremely limited capacity to form normal social relationships in maturity (asocial, autistic, or even sociopathic symptoms may develop [Fox, 1974]).

In conclusion, therefore, it is theorized that the parasympathetic nervous system (and cardiovascular CNS afferents) has evolved from an initially endogenous homeostatic mechanism to an exogenous, socially dependent mechanism is more gregarious and sociable mammals. This "social (evokable) homeostasis" or ethostasis forms the basis for the development of affectional bonds between animals and between animal and man.*

Summary

Several studies of cardiac activity in wild and domesticated canids in various controlled conditions were described. Since changes in heart rate are indicative of sympathetic/parasympathetic activity, monitoring of cardiac activity provides a sensitive, quantifiable index of autonomic functioning and related emotional reactions and overt behavior. Developmental changes in heart rate were studied in several breeds of dog; with increasing age, between 2 and 3 weeks, a gradual slowing in rate was found. Treatment with atropine demonstrated that this was due to the later development of vagal inhibition.

Evidence was presented to support the hypothesis that heart rate, along with other measures of emotionality and behavior, are

---

*In man, and possibly in other mammals, this socially linked and integrated psychophysiological mechanism may be the basis for empathy and altruism. (Lacey and Lacey, 1970, for example, describe one experiment where immersion of the hand into cold water evokes bradycardia; a person observing another doing this will also show a transient bradycardia which is greater if he has recently immersed his own hand into cold water.) Parasympathetic control (corticovisceral feedback) is an essential component of meditation and certain Yogic practices; the physical and psychological benefits from such "mind–body control" for man are now being objectively measured, so far with extremely promising results (Wallace, 1972).

valid indices of temperament. These factors, which as a multifactorial constellation make up what is generally termed temperament, were correlated with individual susceptibility or resistance to physical and psychological stress and disease.

To what degree such factors, and temperament per se, are inborn or innately predetermined, was partially answered by studying individual differences within litters. Environmental influences, especially stress early in life, were shown to have a significant effect upon temperament, including heart rate and other dependent measures such as plasma cortisol levels, dominance, and exploratory behavior.

Following the findings that those canids with high resting heart rates (either innately or as a consequence of experimental manipulation) were superior animals in terms of exploratory behavior, learning ability, and resistance to stress, a large survey of over 100 puppies was conducted. The heart rate was used as one of several indices of temperament evaluation and was found to be a valuable but not exclusive prognostic measure of later potentialities for training such dogs for military work (guarding and locating concealed mines).

Using biotelemetry, contact and petting was shown to evoke a clear bradycardia in dogs. This phenomenon was reliably evoked in dogs socialized to the handler. Bradycardia was shown to occur also following physical contact or with eye contact in wild, unsocialized canids. The psychophysiology of these same reactions occurring paradoxically in such very different contexts was discussed in relation to parasympathetic–sympathetic functions.

These observations, together with the data on individual differences in cardiac activity and effects of early stress experience, were integrated with a relevant theory of autonomic tuning, elaborated by Gellhorn, to account for individual differences in temperament, reactivity, and stress susceptibility in man. The possibility that domestication has contributed to the increased parasympathetic (vagal) tone was explored; individual differences in cardiac activity as an indicator of emotional instability and susceptibility to stress were emphasized to be a valid subject for further study and clinical evaluation.

XI

Domestication and Man–Dog Relationships

In the foregoing material, some of the complexities and variables involved in domestication is discussed. Of special interest for future research, in both farm animal husbandry and in raising pet animals, is the applicability of early handling; such programmed rearing methods are no less relevant in laboratory animal science (Fox, 1971c).

The research studies reported herein add more to our understanding of the temporal organization of behavior and of its inheritance in wild and domestic canids and their hybrids. More research is needed on the subtle interindividual variations in canid vocalizations, although the study reported here does again open up the question as to the origin(s) of the domesticated dog and the possible effects of neoteny and dependency on vocal and other behaviors. Domesticated dogs that have become feral is a remarkable statement for the reversibility of the subtle effects of domestication and of the adaptability of the domestic dog. Experience with hand-raised wild canids adds to the understanding of the role of socio-ecological preadaptations discussed in Chapter 2, and the genetic or species limitations for domestication. This concluding summary helps clarify and integrate the aforementioned material into a more concise appraisal of the effects of domestication and human relationships on canine behavior.

Miscellaneous Aspects of Domestication

It should be emphasized that movements and display postures are more stable under domestication than other characteristics (e.g., display structures); thus, there may be little interference with courtship, mating and other social behaviors. How an animal communicates and displays may be less affected by domestication than when and to whom it will display and when and with whom it will interact. Interactions between different species may affect the amplitude and frequency of action patterns rather than alter the patterns of behavior per se, although in some instances, provided with the right contingencies of reinforcement, new behaviors within the animal's repertoire but not normally manifest may emerge, e.g., canid mimicry of the human grin (Fox, 1971b) and the acquisition of American deaf-dumb sign language in chimpanzees.

Evolutionary changes may be compressed and mimicked in domestication, genetic mutation, selection and migration, and isolation of populations under human influence occurring. Geographic and experiential factors (e.g., sexual imprinting) may enhance or reduce isolation in developing new hybrids and strains. Scott (1968a) succinctly states for the dog that

> The dog is... not in any large degree a conscious product of human ingenuity. Rather it has evolved under the influence of countless thousands of interactions with human masters. We can therefore think of the dog as a species which, on domestication, entered a new habitat and underwent a process of adaptive radiation similar to that of a wild species entering a vacant ecological niche. It subsequently underwent further modification and diversification as it became divided into small local populations and selection pressure became relaxed in certain directions and increased in others under the influence of the human social environment.

The domestic environment can also induce a new phenotypic expression of the same (wild) genotype. Abnormal behavior and genetic instabilities may rise, as in the regression to a wild temperament, in overmagnification of a wild trait by a change in threshold, or in disruption of temporal sequences such as sexual or maternal behavior. Examples include, respectively, shyness, fear of strangers, and neophobia which develops in some pups around 10–12 weeks of age, as in the wolf cub normally; hyperaggressivity and proximity intolerance in certain breeds; aggressive dominance disrupting mating and prey-stalking and killing appearing in bitches with pups. Repression of some natural behaviors may also occur, e.g., normal regurgitation of food for pups is often absent in purebred bitches and parental behavior incomplete or absent in male dogs, since the latter, kept for generations without contact with mother and pups, may show little interest in them and may even compete with and kill them. Many questions remain unanswered, notably why the apparent high mutagenic potential of phenodeviant forms in the dog in contrast to the domesticated cat, or is this mainly due to the dog having been domesticated some 4000 to 6000 years longer and the cat having been kept mainly for one role (to catch vermin) and not selectively developed to serve different roles as is the case in the various breed-classes of dogs. The question of the influence of coat color on temperament is also an intriguing topic, since selection for coat color may influence behavior and emotional reactivity. Trumler (1973), for example, notes that black Alsatians are "livelier" than lighter colored dogs; black is recessive, and some fawn dogs may have black temperament and carry the black gene since if two of the fawn color with black temperament are bred they will produce black offspring. Keeler (1975) and Guthrie (1975) have also discussed the possible significance of melanism in color-phase variants of the red fox in relation to temperament, docility, and resistance to social stress.

A multifactorial analysis of behavioral and physical data from over 60 captive wolves was conducted by one of my graduate students, Randall Lockwood (1976). Interestingly, he found a correlation between behavior and coat color. Adult black color-phase wolves showed less territorial marking and more active and passive submission than brown or tan color-phase wolves.

Trumler refers to "Hecks' law"—when a wild and domestic species are crossed, there is not always a 50:50 distribution of respective parent traits in the hybrids, but rather "qualities came to light which were those of the prototype of the domestic animal." Precisely what this statement implies is difficult to interpret in the absence of further data, qualification of the concept of prototype, and references to increased variance in F_2 hybrids due to independent assortment of Mendelian units. This latter phenomenon is surely a key to increasing diversity and selection of phenodeviants in early domestication. Trumler may be confirming this and Hecks' law with what Konrad Lorenz (1970) demonstrated several years ago in his studies of different species of duck and their hybrids. He found that a hybrid of two species may show certain behaviors not present in either parent species but present in another, possibly more ancient or earlier evolved species. This is termed reactivation, a phenomenological construct which may be interpreted in terms of threshold and temporal sequencing of behavioral units, where omission, reorganization, or repetition of various units may occur as a consequence of either natural evolution (speciation) or domestication.

Some of the specific and potential changes attributable to domestication in the dog are reviewed in this book. The changes fall into a number of more general categories relevant to the effects of domestication in other species as well.

First, the changes may be genetic or environmental in origin, or they may be a consequence of more complex genotype–environment interactions (Fox, 1970b). Second, it is important to define the level of behavioral analysis from which conclusions and inferences are drawn; for example, certain consummatory acts (fixed action patterns) such as those associated with ingestive, aggressive, and sexual behavior may remain unchanged, while their threshold, amplitude, and frequency may be increased or decreased. Also the temporal organization or sequencing of action patterns may be modified (e.g., truncated at various junctions as in prey-catching and killing in canids) as well as the type of stimulus or release which evokes such behavior (e.g., a moving vehicle may become substitute prey for the dog or a person may release sexual responses in an imprinted bird). Generalizations, therefore, cannot be made as to the direction of changes in behavior (see Figure 1).

Genetic influences are generally attributable to inbreeding, directed selection, and genetic drift. The removal of natural selection pressures and of such natural variables as seasonal changes in light, temperature, and availability of food can evoke rapid physiological changes (such as the multiple-estrus of wild felines in captivity, which in the natural state are monoestrous). Structural

ONTOGENY OF CANINE ETHOGRAM

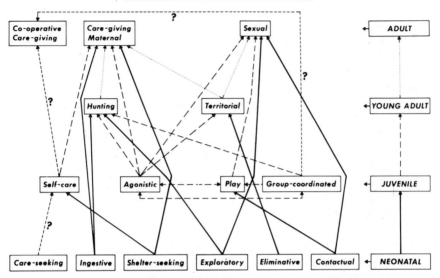

Figure 1.   *Ethogram of the dog (from Fox, 1971a) showing interrelationships in 15 major categories of behavior. Domestication may
influence the intensity (frequency/amplitude and threshold) of these
behaviors and also their interrelationships. Self-care, in an overdependent dog, may be reduced, with related interference in maternal
behavior; hunting behavior may be redirected toward offspring and
disrupt maternal behavior. Reduced agonistic behavior may interfere with sexual behavior, while intensified aggressivity may influence group-coordinated (pack) behavior. This schema therefore de-
picts the interrelationships among various categories of behavior,
which must be evaluated in order to determine effects of domestication in a given species or breed. Analysis at the molecular level is
also essential. The temporal organization and sequencing, stimulus
control (releasers), response threshold, frequency and amplitude of
associated action patterns with each behavioral category, such as in
hunting or maternal behavior, should be determined (see text for
examples and further details).*

changes, especially in display structures and social releasers
(horns, bumps, body coloration, and specific social marks) may be
brought about slowly through selective breeding or more rapidly
by husbandry procedures such as castration. All these variables,

together with the social composition of the captive animals (kept together in natural groups, free-ranging, or confined in groups of the same sex and age), can produce further changes in social behavior, growth rate, sexual behavior, etc. Thus, the method of animal husbandry or mode of livestock management can greatly influence the animal's phenotype and presumably, over time and generations, the genotype as well.

In an interesting experiment with rats, Hughes (1973)* evaluated three major environmental variables which may effect docility, namely, cross-fostering (of wild offspring onto a domesticated mother), rearing in an enriched environment (seminatural, complex cage as distinct from the small barren cage of standardized laboratory animal care), and early handling or gentling of the wild offspring. As might be anticipated, only the latter procedure had any significant effect in enhancing docility.

Thus in order to evaluate the effects of domestication, the environment which the species is exposed to and is raised in must be known and operationally defined in terms of its actual and potential influences upon behavior, structure, and endocrine functions. Since the domestic environment is often unnatural to an extreme, abnormal yet adaptive changes in behavior, structure, and function may develop through phenotypic change or genotypic selection; atypical strains or genotypes may then emerge, quite different from the original wild form (see Figure 2, Chapter 9).

Domestication may generally increase genetic variability and consequently lead to a greater degree of phenotypic plasticity, since the more unidirectional pressures of natural selection have been eliminated to varying degrees. Some researchers also believe that domestication causes a general degeneration, both physically and psychologically. This may be true of certain systems or structures in some species, but as emphasized earlier, no single generalization as to the direction of the effect of domestication can be made. Degeneration, for example, may be extremely difficult to distinguish from infantilism or neoteny (paedomorphosis).

A thorough analysis of domestication effects in any single species must not only take into account the genetic, environmen-

---

*Hughes, C. W. (1973). *Early Experience in Domestication.* Unpublished doctoral dissertation, University of Missouri, Columbia, Missouri.

tal, and experiential variables discussed earlier, it must also encounter the limitations of comparative studies with a wild model
ancestor. The latter, adapted to captivity for study purposes, may
differ in many ways from one born and raised in the wild, for
which it is preadapted. Similarly, the domesticated species may
differ from one that is raised under seminatural conditions or has
actually become feral. The conscious or coincidental changes
wrought by man's intervention must also be determined, for
example, rigorous selection for docility may increase proximity tolerance but may negatively influence fertility. Similarly, paradoxical
selection may lead to behavioral problems, as in selecting Alsatians
to be highly sensitive, alert, and reactive; a number become
hyperactive, overreactive, and extremely timid; also selecting and
training a dog to be a protector or guard of person or property may
set up a conflict with the demands to be sociable and friendly at
other times. Such coincidental or unidirected changes and
paradoxical sequelae to consciously directed selection and training
for specific utilitarian purposes must be identified before a more
complete understanding of the effects of domestication can be
gained.

Origin of the Dog

The origin of the domestic dog, however, is still an enigma. Although it is generally held that man initially domesticated the dog
because of its value as a hunting partner, others, like Meggitt
(1965), cast doubt on this. His studies of the association between
Australian aborigines and dingos show that the dingo is a virtually
untrainable canid, even when hand-raised. Its main value is to
keep the camp clean of garbage (and human excrement), to keep its
human companions warm at night, and occasionally to act as a
watchdog, giving warning of intruders. They are effective as
cooperative hunters only in tropical rain forests, a fact supported
by the use of the Basenji in Africa for hunting in a similar rain-
forest environment. Presumably therefore, the early stages of
domestication may not have involved the use of the dog's hunting
abilities except in suitable environments and only later, with selec-

tive breeding, did the dog come to fulfill a hunting role. Initially, it would seem that the dog was a camp companion, follower, and guard, and later its roles as hunter, draft animal, protector, and herder of livestock were developed. Domestication of the dog took place while man was still in the hunter-gatherer stage according to Dr. Barbara Lawrence* who has found remains of dog in North America dated 8400 B.C., which indicates that the dog was the earliest of man's domestic animals. The dog later became firmly established in village farming communities, at which time, presumably, selective breeding for particular utilitarian functions was undertaken.

Epstein (1971) has presented the most extensive and detailed review of the origins and descent of the pariah and other types of dog, with particular emphasis on those of Africa. He reaches the conclusion, based upon extensive critical review of archaeological reports, that the small southern asiatic wolf, C. lupus pallipes, is most probably the dog's ancestor, discounting jackal ancestry on the basis of chromosome number (jackal, 74; dog and wolf, 78). However, Chiarelli (1975) reports 78 chromosomes for the coyote, wolf, golden jackal, Cape hunting dog, and domestic dog; chromosome numbers are therefore of questionable significance.

Epstein (1971) concludes that

Initially the domesticated dog was of no economic use to man, save for its doubtful role as a scavenger. Only gradually, and in the course of its diffusion, it became a utility animal—an object of sacrifice, of ceremonial or profane consumption, or employed as a guardian of the home and flocks, in the hunt, for draught and for its wool, or merely tolerated as a scavenging pariah. It may be assumed that it was only after realisation of the usefulness of the first domesticated species that the idea and practise of domestication were transferred to other species.

While this is a plausible hypothesis which I would support, it must surely refute his notion (and the widely held view) of wolf ancestry, since the latter would, because of its size and social behavior as a pack animal, be an unlikely canid candidate for camp/settlement scavenger. Small contemporary subspecies of wolves in Southern

*Museum of Comparative Zoology, Harvard University.

Europe and Asia have adopted a more jackal-like mode of a solitary scavenger because of shortage of available prey. It is improbable that early man lived in regions where prey was so scarce, but rather both the wolf pack and the human hunting-gathering group shared a comparable niche rich in available game. The role of scavenger would then be assumed by jackal- and dingo-like canids in association with human campsites and following wolf-pack hunts.

With the advent of agricultural settlements and increasing destruction of natural habitat, some wolves may have become scavengers close to such settlements, while others moved further away into more remote uninhabited regions. The former would have been a threat to the livestock that were domesticated at a later date and would have certainly been discouraged. Jackals, however, are smaller in size and less powerful than wolves and would have been tolerated more since they would be little threat to livestock or people. In a parasitic/symbiotic mode comparable to contemporary jackals and feral pariah dogs, the primitive prototype of the domesticated dog was undoubtedly tolerated by early man [as suggested by Epstein (1971)], and this prototype was unlikely to have been wolf-like. Therefore, the logical conclusion is that the domesticated dog arose from a more primitive prototype (possibly dingo-like) which evolved from the wolf/jackal stem prior to man's intervention and prior to its initial association with man.

One must seriously question Scott's (1968) notion of the affinity between dog and wolf in terms of high sociability and pack formation (which, he argues, is strong evidence to support the view that the wolf is the ancestor of the domesticated dog). First, environmental influences, especially the availability, distribution, and type of prey affect sociability and pack formation (Fox, 1975a). For example, in Italy, wolves tend to be solitary and have adopted a scavenging mode of existence similar to the jackal and coyote. The latter two species, normally living solitary lives or in pairs, may be found in packs and in areas where food is plentiful. Thus, environment influences sociability, and in the dog there is a comparable picture complicated by man's influence. This influence, keeping a dog on one's property, leads to territoriality and has a socifugal or dispersing effect, while free-roaming and homeless dogs in the same locale may establish a dominance hierarchy and hunt in a pack, provided food is plentiful. This was observed on a recent field trip to Southwest India where a study of village pariah or pie-dogs revealed this environmental influence on social behavior.

Studies of Pariah Dogs

Observations of home-owned free-roaming and feral pariah dogs in India revealed the complexity of their social organization in the typical rural village environment. Adopting and socializing one adult female and following her interactions with neighboring dogs added further insights in support of the following tentative generalizations.

There are basically three types of pariah dog: (1) home-owned and not inclined to roam more than a few hundred yards from home, (2) home-owned but free-roaming, (3) ownerless and free-roaming, with or without its own home-base territory, a den, or shelter. Gregariousness increases from (1)–(3), attachment to man having an intraspecific socifugal effect. In (1) the dog often manifests relative dominance (Leyhausen, 1973) in that within its home-base territory it is dominant over intruders. In (1) its foraging/hunting range may be restricted by two factors, namely, if it is given food by its owner (which is rare) and if it is subordinate to neighboring dogs. Foraging and hunting ranges overlap, and consequently, where dogs meet away from their home bases, absolute dominance is seen. While a dog may drive a neighbor off its (owner's) property, it may be subordinate to the same dog on the neutral territory of a shared foraging/hunting range. Thus, attachment to man and place has a socifugal effect, while sharing of the same food area leads to the more typical dominance–subordinate hierarchy. Here, competition over available food concentrated in a limited area increases conflict, and an absolute dominance hierarchy is established. Interestingly, though, this constitutes a loose social unit (not a pack per se) and pariah dogs of types (1) and (2) especially will temporarily pack together to hunt deer (an activity encouraged by villagers) and to drive off a strange dog that enters their shared range or collective territory.

   Different relationships between the sexes, between dogs of different ages, and females in heat and with pups also influence the type of interaction between conspecifics of all three classes, ranging from active and passive submission to indifference or overt aggression.

   Type (3) dogs may aggregate more frequently and form temporary packs, sleeping, foraging, and hunting together, since they lack the socifugal influence of having an owner and home-base territory to defend. In larger villages, where several sources of food may be present (butcher and baker stalls), dogs may set up a territory around a source and keep others away. Many such dogs are homeless and belong to type (3), but competition over the focal food source sets up a territorial situation analogous to the home-based dogs of type (1) and (2). In one village, there was clearly one pariah dog that was the most dominant dog in the community, having unchallenged freedom of movement and access to all food sources. In summary, the influence of man in providing one or more dogs with a home (shelter with or without food) introduces socifugal territoriality which is complicated further by the fact that dogs may compete or cooperate in foraging and hunting and in defending their shared range or collective territory from nonresident free-roaming and feral dogs. With a shared range, dogs of all three types may be involved. The capacity to form a pack, based upon a social dominance hierarchy, is limited by these socifugal factors, but even so, this loose social unit of neighboring dogs can form a temporary pack in a collective defense of shared foraging range/territory against intruders and in hunting large prey in the forest or jungle close to the village. Subsequently, though, the pack dissolves into a loosely structured social unit by virtue of the fact that some dogs have home territories to return to. Also, the supply and distribution of their usual source of food deters pack formation, since they must forage alone or in small numbers in order to minimize competition and conflict and optimize the utilization of such resources (see Figure 2).

   From their extensive studies of the pariah dog, Menzel and Menzel (1948) recognize five basic types: heavy extreme (sheep dog like), heavy medium type (dingo-like), light medium type (collie-like), light extreme type (greyhound-like) and a small-grown type (toy dog-like). They note that pariah dogs may live in packs or

Figure 2.   *Indian village pariah dogs showing a typically emaciated bitch nursing pups and males forming a temporary pack around a female in heat. Village dogs tend to establish territorial zones around local food sources but in the absence of such may form mixed nomadic packs.*

alone. In the latter, where dogs share the same home range with others, there is some form of social organization as a loose unit, since they will band together and drive off strange dogs, as I also observed in India. Presumably the interrelationships among domesticated dogs with homes, homeless village-resident scavengers, and more peripheral feral dogs are a complex interplay of territoriality, competition/cooperation over food (garbage and wild prey), with each other (and with wild species such as jackals), and social dominance–subordinate relationships relative also to the proximity of home base or natural den/shelter (i.e., the center of each animal's territory). (Territorial zones may overlap and several dogs may share the same hunting/scavenging range). Sociability will vary, therefore, not only in terms of the individual's temperament but also in relation to these socio-ecological variables of space (place) and availability and distribution of food. Since pariah dogs may live singly or in packs, the flexibility in their social behavior points more to environmental adaptations than to support of the view of wolf ancestry as proposed by Scott (1968a). Menzel and Menzel (1948) conclude that

> The question of the pariah dog is among the most interesting of zoological problems, particularly from the point of view of racial history. We can be quite sure that an examination of the history of the domestic dog in all its aspects would reveal that there is scarcely a single facet that might not be illuminated by research into the problem of the pariah dog. . . .
>
> In spite of varying theories on the descent of the dog, most authors recognize two distinct sub-groups of the genus in the old world which differ from each other in racial history—the *Northern* and the *Southern Dogs*. The pariah dog belongs to the Southern group. Studer does not race the Southern dogs (like those in the Northern group) back to the wolf, but to another kind of *Canides*. He writes as follows: "From the time of the Flood there existed a type of *Canides* (*Canis ferus*) smaller than the wolf, distributed in the same area but exceeding it in number in the South, consequently spreading as far as the Australian continent. This kind is divided into two main varieties . . . in the South it is represented by the dingo."

The pariah dog is a prototype perhaps of the domesticated dog, living today in India under conditions analogous to those of early agricultural man in small village communities. It shows a high degree of social flexibility—solitary to pack formation—which in no way supports or negates a wolf ancestry. No conclusions pro or con can be drawn of wolf ancestry from studying the pariah dog, but rather, they manifest a degree of social flexibility which is adaptive to the environment and which is neither unique to dog or wolf. Under certain conditions (Fox, 1975), coyote and jackal may form packs from a loose social unit, and wolf packs may break up and follow an adaptive social model typical of the more solitary coyote and jackal.

Studies of free-roaming and feral urban and rural dogs in the United States similarly show that with abundance of food (prey in rural areas) dogs will form packs, while in urban areas where food is less plentiful, packs are rare and a solitary mode of foraging usually disrupts pack formation with a few exceptions. Dogs with homes (and usually, therefore, a constant source of food) do not generally roam far; they tend to be territorial rather than running in neighborhood packs and the latter, when seen, are usually merely temporary groups of males around a female in estrus as detailed in Chapter 3. Thus, the contention that since dogs are highly social like the pack-forming wolf, the wolf must be the ancestor of the dog, is untenable.

Three other factors supporting wolf ancestry must also be questioned. Scott (1968a) states that dog and wolf have very similar behavior patterns and vocalizations. The latter is certainly not true; the range and variety of howls recorded from a single wolf contrasts the limited range of howls in domesticated dogs, as well as other vocal characteristics discussed earlier.

Behavior patterns are very similar in dog and wolf, there being a closer affinity between these two species than between dog and coyote or jackal and other canids reclassified in terms of behavioral rather than structural similarities (see Figure 1, Chapter 2). While this evidence may lead to the conclusion that the dog is a domesticated wolf, it equally implies that they may share a common ancestry prior to domestication and that the dog was a dog before it was domesticated. This view is supported by the fact that behavior patterns per se are influenced little by domestication: how an animal behaves is more or less phylogenetically fixed, but when and to whom it behaves as well as the threshold and sequencing of behavioral units is affected by domestication and early experience. Structure is more influenced by domestication than behavior; a bulldog, Irish wolfhound, Chihuahua, and dachshund, structurally very different, share the same basic behavioral repertoire.

The change in structure which most archaeologists and taxonomists recognize as a consequence of domestication is the reduction in tooth size; even a large breed like the Irish wolfhound has small teeth relative to its skull size and in comparison to the wolf. This may, however, simply indicate that this breed type is a giant mutant from a medium type having a tooth size more comparable to a large jackal or small wolf. So a small wolf has been sought as the probable ancestor of the dog, and the Asiatic wolf, Canis lupus pallipes, is now accepted by most taxonomists and canid ethologists as the dog's ancestor. Yet there is no evidence of a transitional form; all that excavations have revealed in the Middle East to date are remains of a small wolf, large and small species of jackal, and what is thought to be dog, with skull and jaw fragments which overlap in measurements with the former two species and so make identification difficult (Clutton-Brock, 1969). The contemporary native pariah dog is very similar structurally to these 8,000- to 10,000-year-old canid remains. Berry (1969), in an important paper relative to this archaeological dilemma, points out that natural environmental changes can mimic changes attributed to domestication, e.g., a reduction in tooth and jaw size may be misidentified in the case of the wolf as being due to domestication. Pre-Pleistocene wolves were of much larger build in the Middle East than the existing subspecies today.

There is a cline of blotched tabby in European wild cats similar to the coat pattern thought to be an exclusive trait of domestic cats. Berry concludes that

> even for a single species, it may be extremely difficult to lay down criteria to distinguish between domesticated and non-domesticated forms . . . there is no reason to believe that domestication per se will alter the phenotype or even a major part of the genotype.

Further review of the earliest remains of what has been identified as *C. familiaris* gives dates of 14,000–10,000 years B.C. as the earliest date of domestication. Interestingly, Macintosh (1975) reports that the Australian aborigines preceded the dingo by 20,000 years, since the first signs of dingo remains are dated around 10,000–8,000 B.C. Origin of dingo and domesticated dog alike remain an enigma archaeologically.

General (but unfounded) consensus holds that when it was first domesticated about 10,000 years ago, the dog was derived from a wolf, possibly the Asiatic wolf (*Canis lupus pallipes*). Archaeological records, however, show that a dingo-like canid was widespread throughout Europe in the Stone Age (Macintosh, 1975). It is my contention that the domestic dog was derived primarily from this prototype of *Canis familiaris*, and at different places and times this canid was crossbred with indigenous wolves to produce some of the more wolfish breeds, such as the malemute and the husky. Although the dingo in Australia is somewhat like the coyote in its social behavior, being relatively solitary, living in pairs, and not forming packs, it is the ecology that has determined this kind of social behavior. If prey were more abundant, pack formation might be possible. Today, the domestic dog is capable of reverting to the wild (becoming feral) and surviving well, either singly, as a scavenger or opportunistic hunter, or combining with other dogs and forming a pack. The pack instinct of the domestic dog and its responsiveness toward a pack leader may well be the basis for its integration into the human family. In other words, a transference of the pack relationship to a family allegiance occurs, and this natural instinct greatly facilitates domestication and training. In fact, if socialization is delayed, the dog is virtually untrainable.

Behavioral Effects of Domestication

Further behavioral evidence of wolf ancestry from Scott (1973) must also be challenged. He suggests that the muzzle-bite greeting ritual of the wolf (detailed by Fox, 1971b as a species-typical action) was selectively eliminated in domesticating the dog since it is frightening to man. Having kept wolves myself and knowing others who have raised them and who are familiar with their behavior, there is no anxiety around such interaction. Also some northern dog breeds which probably do have recent wolf origin (malemute and husky) show this muzzle-biting behavior and breeders have no concern or anxiety about this trait.

Scott's recognition of the different tail carriage of dog versus wolf is important, but his interpretation is questionable. Dogs, he observes, have a sickle-shaped or curly tail. He interprets this in the dog as a mutation; a single mutation suggesting that until there is further evidence to the contrary, the dog was probably domesticated as a mutant form from the wolf only once in one place. This tail mutation was, he believes, "preserved because it was useful in distinguishing wild from domestic animals"; early man could surely recognize a socialized canid independent of its tail! This attractive conclusion is unwarranted however. It is not so much the structure and carriage of the tail that is significant (even though the Spitz-type dogs always carry it high) but the differences in tail displays in dog and wolf. If it is accepted that structure (curly or docked tail) changes more under the influence of domestication than behavior (tail postures in communication), then the wolf ancestry of the dog seems even more remote. For example, in the threat posture of the wolf, the tail is held vertically erect at an angle of 75°–85° from the horizontal (of the back), while in the dog it is curved as an arch over the back, even in those breeds (and eastern pariah dogs) that carry the tail low in the same position as the wolf when at rest (see Figure 3).

Having negated all existing behavioral evidence that purports wolf ancestry to the domesticated dog and seriously questioned supporting archaeological data, the enigma of the origin of the dog remains unsolved. Lorenz (1975), who has refuted his earlier speculation of a dual wolf-jackal dog ancestry, has now fallen in with the general consensus in favor of wolf ancestry, again, as with the case for the jackal, with no clear evidence. Menzel and Menzel (1948), who studied contemporary pariah dogs in some detail, point out the dilemma that one cannot today identify a wild type from the domestic since the latter may become feral and live independent of man. Dogs may also hybridize with wolf, jackal, and coyote, but presumably ethological and ecological barriers in the wild usually prevent this. The point is that there is no pure wild form of *C. familiaris* in the wild today other than feral packs which, like a wild species, will eventually breed to a high degree of phenotypic uniformity after a few generations.

Another clear distinction between dog and wolf is the supracaudal tail gland (Fox, 1971b). While the gland, marked by a stripe of dark hair on the upper surface of the tail, is active in the wolf, no odor can usually be detected in the dog.* Furthermore, if there is a distinctive hair mark over the gland in the domestic dog, the mark (of darker or thicker hair) is usually triangular as in the coyote and jackal and not a thin stripe as in the wolf. This is surely

---

*When an odor can be detected, it is, to the human nose, distinctively "doggy" in contrast to the sweet-mown hay odor of the wolf.

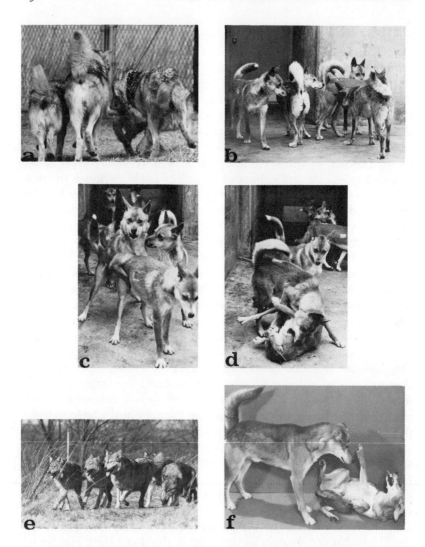

Figure 3. *Distinctive traits of wolf and dingo. (a) Vertical tail in dominant wolf (in center); (b) arched tail in dominant dingos—all asserting high rank; (c) and (d) rank conflict between dingos on introduction, including aggressive clasping and (inhibited) cheek-oriented bite; (e) in same period together wolves form an integrated pack in contrast to more individualistic dingos; (f) dominance-submission in F_2 coyote × dogs; note doglike arched tail position.*

strong evidence against recent wolf ancestry in the dog. (See Table I for a general overview of wolf/dog differences.)

Comparative studies on the behavior and development of wild canids has enabled us to identify a number of changes, which in the domestic dog may be attributed to the effects of domestication. For example, the wolf reaches sexual maturity at 2 years of age, while other wild canids attain full sexual maturity at 1 year of age. In the wolf, this delayed sexual maturation might have important social consequences, in that the young will stay around for an extra year to help adults raise another litter. In the domestic dog, sexual maturity is reached at a much earlier age (6–9 months). Domestication, therefore, has accelerated the development of the endocrine glands associated with reproduction. Related to the attainment of full sexual activity in the wild canids is the emergence of aggression, dominance relationships, and defense of territory. Interestingly, in the domestic dog, these behaviors appear on the normal time base at 12–18 months of age. This suggests that the more central emotional aspects related to full sexual maturity have been split from the endocrine aspects of reproductive behavior through domestication. For example, by 1 year of age a coyote has attained sexual maturity and at the same time will defend its territory against intruders and will engage in dominance and male rivalry fights. In the domestic dog full reproductive potential is attained much earlier in life, while problems associated with aggression toward intruders and dominance fights with the owner do not develop until a later age. The males of wild canids produce sperm seasonally, while in the domestic dog the male is constantly fertile. Whereas a wolf may show specific mate preferences, such a trait in the dog may have been eliminated, since this would make selective breeding under the choice of the owner a difficult task if the stud or bitch refused to mate with a selected animal.

Comparative studies of coyotes, wolves, beagles, and other domesticated breeds, all raised under identical conditions, have shown that on separation the domestic dogs show much more distress. It can be concluded from these observations that the domestic dog is basically much more dependent than the wild species. The greater the dependency, the easier it is to train such an animal. Hand-raised wolves and coyotes, though extremely sociable and friendly toward the handler, are notoriously difficult to train. It may be proposed, therefore, that another consequence of

Table I.
SOME EFFECTS OF DOMESTICATION: COMPARISONS OF WOLF AND DOG

Trait	Wolf	Dog
General activity level	high, especially if high-ranking	great breed variation—hypo-hyperactive
Exploratory (investigatory) behavior	high, especially if high-ranking	low-high, some "specialized" in one sensory modality
Neophobia (fear of strangers/unfamiliar objects)	common, usually low habituation	undesirable trait selected against—e.g., touch-shy, 'spooky' etc:: rare, high tolerance and habituation.
Secondary socialization (social potential)	limited	enhanced in most breeds except guard dogs
Dependence/trainability	minimal	enhanced in most breeds
Vocalizations		
group howling[a]	common social activity	similar, but includes barks and yelps
barking[a]	rare—threat only	common in many contexts
yelping[a]	absent	common in many contexts
Displays and action patterns	complex and variable repertoire	basically similar to wolf
muzzle-biting and pinning[a]	common—ritual display of dominance	absent except in wolflike breeds, e.g., malemute

aggressive hip-slam	common	rare except in wolflike breeds
vertical tail in threat[a]	usual display in dominant wolf	absent—tail arched in threat
face-lick greeting	common, but low frequency	higher frequency, especially in toy breeds
genital grooming	a social activity	absent
leap-leap play action[a]	absent—bounding run at conspecifics	common action in dogs and dog × wolf hybrids
Sequences of behavior, e.g., hunting-tracking, herding, stalking, killing and retrieving	clear temporal sequence, with social cooperation and hunting strategy, e.g., waiting in ambush: flexible senses	full sequence in some; inhibited or truncated sequences in "specialist" breeds, e.g., sheep dog, pointer, retriever: also specialized use of senses in some breeds
Allelomimetic (group-coordinated "packing" behavior	intense, plus strong following response to leader	variable—low in terriers, high in pack hounds
Social (dominance) hierarchy	complex, but basically linear	common, but absent in some breeds in groups, e.g., beagle
Intraspecific aggression and proximity intolerance	varies according to rank	greater extreme variations with breed
male-male	moderate	variable according to breed
female-female	high in breeding season	variable according to breed; higher in season

Table I. (*continued*)

Sexual behavior		
maturation	2 years	6–9 months
female season	one annual estrus	two heats per year
male season	seasonal spermatogenesis	constant spermatogenesis
mate preference	dominant female's choice	often indiscriminate
marking behavior	greater in dominant male and female	very high in all males, associated with "tonic" sexuality
Maternal/paternal behavior	male and female regurgitate for cubs	variable, often absent especially in males
Major physical characteristics		
teeth	well-developed carnassials, canines and "caniform" upper outer incisor	all reduced, also premolars—some often absent
coat color	usually adaptively cryptic and unbroken	great variation in color and texture
underfur	dense in winter	absent in some or reduced, or reversed—guard hairs absent, e.g., poodle
seasonal moult	late spring	often seasonal or disrupted
supracaudal tail gland[a]	present	inactive except in wolflike breeds, e.g., malemute
brain size (relative to body proportion)	—	reduced up to 30%

[a]Wolf–dog differences may imply separate ancestry rather than effects of domestication per se.

domestication has been to make the domestic dog extremely dependent. The degree of dependency varies from breed to breed. For example, malemutes and huskies are aloof and are much more difficult to train than the more dependent and willing terriers and toy breeds. In the toy breeds, there is not only a greater dependency but also a neotenization or infantilism, whereby the physical structure of the adult animal more resembles an infant. Tied in with the relative immaturity of the structure is the perpetuation of infantile behaviors into adulthood. How much of this is genetic and how much is related to the way in which the owner raises the animal is open to question. The capacity to retain infantile behaviors into maturity must be genetic.* Comparative studies with the wild canids reveal that the wolf has considerably more infantile behaviors (which are called derived infantile actions) as an adult, compared to other wild species. This, coupled with a strong tendency to follow superior animals, helps in the integration and coordination of pack activity. Infantilism and responsiveness to a leader or parent figure is no less evident in the domestic dog. However, because of the greater degree of demonstrable dependency in the many breeds of domestic dog, the possibility for developing emotional disorders is enhanced, but at the same time, it is a more fulfilling pet for those who need an extremely dependent relationship.

Man–Dog Relationships and Human Needs†

Human relationships are now changing—young people are not having any children or are delaying having children for a number of years after their marriage. Senior citizens, retired people,

---

*See Fox (1975c, 1976) for further details.

†Although the negative aspect of excessive inbreeding is now well recognized as a major flaw in the propagation of "pure breeds" of domesticated dogs, the perhaps more serious problem of "overbreeding" is not fully recognized (often the former

widows, and widowers whose offspring might now be living many miles away have a need for a close companion, be it cat, dog, fish, or parakeet. It is not only the patterns of human social life that have changed over the past 20 or 30 years, but also human needs as well. The need for companionship, for example, is exaggerated when families are separated and when people are lonely and alienated in a depersonalizing urban environment. The more dependent the pet is for some owners, the more fulfilling it is as a companion or as a child substitute. It is this dependency, however, that opens the doors to a number of psychosomatic and psychogenic emotional disorders, some of which are analogous to those described by child psychiatrists. I believe, therefore, that the incidence of some of these disorders will increase and that the veterinarian in small animal practice will have to be on the lookout for such disorders in the near future, if not today.

Critics might say that to attribute a dog with humanlike emotions and needs is to be unscientific and anthropomorphic. Research has shown, however, that the developing brain of the dog, its unfolding pattern of socialization and other critical and sensitive periods during development are very similar, and sometimes identical, to the same phenomena recognized in the human infant, although they develop on a different time base (Fox, 1971a). The dog has basically the same limbic or emotional structures capable of generating specific feelings or affects reflected in overt emotional reactions and also in changes in sympathetic and parasympathetic activity which are linked with psychosomatic and emotional disorders. Add to this common neural substrate shared by dog and human infant the important variable of attachment which is a consequence of socialization, as between dog and owner and child and

---

being mistaken for the latter). Overbreeding implies undirected selection where man does not replicate natural (directed) selection processes essential for the continuance of normal structural and psychological functions through succeeding generations. For example, natural social selection favors bite inhibition and strong parental tendencies in canids. Without culling or not breeding dogs that lack bite inhibition or have other behavioral anomalies, domestication can increase variance and the frequency of maladaptive types. Gradual degeneration, therefore, is occurring in those breeds of dog (especially if they become popular) where man does not mimic the processes of natural selection whereby normative structural, physiological, and behavioral functions are maintained and stabilized in the species or breed variety over generations.

parent, then it should not be a surprise that both dog and child under certain conditions may develop analogous or homologous behavior disorders. These can range (Fox, 1968, Brunner, 1968) from psychogenic epilepsy to asthma-like conditions, compulsive eating, sympathy lameness, hypermotility of the intestines with hemorrhagic gastroenteritis, possibly ulcerative colitis, not to mention sibling rivalry, extreme jealousy, aggression, depression, and refusal to eat food (anorexia nervosa).

Types of Relationships

The following is an arbitrary classification of various types of relationships which may be established between the owner and the dog (Fox, 1976c): object-oriented; utilitarian–exploitation; need-dependence; transpersonal relatedness; stewardship. The most general one is simply need-dependent companionship. The pet fulfills the social need in the owner for company and vice versa.

Another relationship with a dog is simply one of a utilitarian working relationship where the dog is employed as a guard, as a guide for a blind person, or is used for work such as herding sheep, driving cattle, or for sport, such as a gun dog or foxhound. The use of the dog as a guard is very much on the increase today. People living in suburbia are increasingly paranoid about crime and violence and will buy a dog such as a Doberman pinscher or German shepherd that they will have attack-trained. More recently dogs have been used as canine co-therapists by clinical psychologists utilizing the dog as a therapeutic bridge with the patient (Corson *et al.*, 1975).

Some Consequences of Pet-Owner Relationships

It is the close symbiotic relationship between dog and owner that can be the foundation for a number of emotional and psychosomatic disorders. For example, the dog that is overindulged and is raised literally as a child substitute may develop a variety of behavioral abnormalities when its relationship with the owner is threatened—as by the birth of a child, by the introduction of another pet, cat, or dog into the household, or by the arrival of house guests. Separation from the owner due to the owner being sick or the dog being boarded when the owner goes on vacation can similarly trigger behavioral pathologies. These include unpredictable aggression, depression, anorexia nervosa, hivelike reactions and pruritis, excessive eating, sympathy lameness, convulsions, asthma-like conditions, cardiospasms, vomiting, and intestinal disorders, including hemorrhagic enteritis. The overindulged dog may also be underdisciplined and when it reaches full sexual maturity it will behave like a socially maladjusted "canine delinquent." It may effectively win the dominance fight with its owner, and it may become the overlord of the household. Such socially maladjusted dogs are extremely difficult to handle on their own territory and can be no less difficult to handle in the hospital. It is important for the veterinary surgeon to establish his dominance over such a dog, and it might be discretely done in the absence of the owner.

In making a diagnosis, it is important to look into the family background of the pet, and to be aware of the dynamics in the household and to be alert to any recent changes within the home environment. Some emotional disorders may disappear spontaneously when the animal is hospitalized and is removed from such aggravating circumstances. A careful differential diagnosis must of course be made and the possibility of allergic, organic, and other infectious causes must be considered. The real crux of the problem is that the close symbiotic relationship can be the genesis of a number of dependency disorders in the dog which can be expressed behaviorally or psychologically and somatically.

Fritz Perls, the founder of human Gestalt therapy, and other clinical psychologists observe that it is dependency and the fear of rejection in man that is the cause of most emotional disorders. In contrast to the dog, the cat suffers far fewer emotional disorders. This may well be because the cat is a less dependent species than the dog. In nature, cats are relatively solitary, and domestication so far has had little influence on their social behavior. They have not been neotenized nor have they been made much more dependent than their wild counterpart. The kind of person who will prefer a cat to a dog as a pet is perhaps less likely to need to indulge such an animal (Fox, 1974). I feel that many people who need a dog are dependent, they tend to be other-directed, and they gain considerable emotional satisfaction from having a dependent companion in their lives. However, more independent, inner-directed people will keep a cat in preference to a dog simply for its aesthetic qualities and its less demanding attitude. The breed of dog that a person owns may be a projection of deeper needs and identifications. An insecure or paranoid person may want a powerful guard dog. Another person who is attempting to live up to an ego image of grace and agility may keep an Afghan hound or a Saluki. It is primarily because of these reasons that the pet often resembles the owner—it is something more than mere coincidence.

Against a background of controlled experimental research and a scattering of clinical case histories, we are beginning to understand more completely how domestication and socialization influence the behavior of man's closest companion, the dog. Such awareness will, I hope, not only improve future relationships between pet and owner but also the relationships between human beings in general.

Postscript

Some may demean domesticated animals as being degenerate or inferior forms of their wild ancestors or counterparts. Others may see them as merely utilitarian "tools," man-made to serve humanity, in order to satisfy and gratify our many and diverse needs. Yet do we fully understand our enormous obligation and debt to them, which is ethically far greater perhaps than our debt to wild forms? While the latter may be in our trust and we their stewards, the former are our own creations. Being so, what kind of creator are we, and are we to become? Our debt to them is unmeasureable, for we have learned and are still learning from them to become more fully human: responsible and compassionate. We can learn through them in countless ways about nature and about our own nature as well. If we knew everything there is to know about a hair on a dog's back, we would know everything about the entire world, since everything is interrelated in evolutionary time and global space.

As we gain wisdom, compassion and empathy through them, and learn from our mistakes as they suffer for us, we may more surely become the creative and responsible stewards and co-creators of and for all life.

Appendix I

The handling procedures consisted of 1 hr of stimulation daily, comprising 10 min photic stimulation in a light- and soundproof box with a flashing light stimulus of 0.16 intensity and approximately 1 sec frequency, 10 min labyrinthine stimulation consisting of 5 min anteroposterior and 5 min bitemporal tilting at an approximate frequency of 1.5 sec through an excursion of 45° from the horizontal; 10 min auditory stimulation, 2 min each at 1, 10, 10^2 and 10^3 cycles/sec at an intensity of 1.0 v and duration of 1.0 msec. This was followed by 1 min exposure in a cold room at 37°F; 5 min in a centrifugal rotator at approximately 45 rpm, and 10 min handling, during which time a series of reflexes were evoked, including the Magnus, rooting, righting, geotaxic, pain, and panniculus reflexes (see Fox, 1965 and 1966). By eliciting these reflexes, the rate of reflex development could be assessed and compared in different treatment groups. The subjects were then placed in a water bath at 80°F and were immersed for 15 sec, during which time they would swim; this was done during the first 3 weeks. Subjects were then rubbed dry with a hand towel and groomed with a soft brush for 10 min and received 2 min general cutaneous stimulation with an air jet (60°F). From 3 weeks of age onward, the handling period was increased to include 10 min play with the operator. After this handling period, they were returned to the mother. Control subjects were kept under typical rearing conditions with the mother, having frequent scheduled human contact of twice-daily feeding and cleaning routines. All subjects were weighed and heart rates were recorded at weekly intervals while the pup was lying quietly in the hands of the investigator. The pups were weaned at 4 weeks of age, received gamma globulin and piperazine anthelmintic and were reared singly in metabolism cages in the animal house environment.

At 5 weeks of age, the subjects were tested singly in a behavior arena equipped with one-way windows to enable the experimenter to observe the animals without being seen. The arena contained cloth bedding from the mother of the pups and, in another corner, a brightly colored child's toy. The reactions of the subjects were observed for 5 min when they were first placed in the arena, and the observations continued for 5 min more after these objects had been removed. The subjects were then replaced for 5 min more of observation. The observers with stopwatches independently recorded the duration of certain activities of each pup throughout each of the three 5-min observation periods, and time scores were then averaged for each group. The activities observed and time-scored were as follows:

1. *Specific Interaction with Stimulus.* Duration of interaction with either cloth or toy, including approach, play, chewing, licking, carrying, and lying beside or running around the object, was recorded.

2. *Nonspecific Exploratory Activity.* The time spent exploring the arena was recorded, including sniffing and licking walls and floor and jumping up at walls, looking up at walls (visual), and attention to extraneous noises (air conditioner turned on as sound blanket). During this activity period, pup never approached cloth or toy.

3. *Random Activity.* The time spent sitting or pacing the arena without any overt reaction to cloth or toy or attention toward walls, floor, or extraneous noises was recorded.

4. *Distress Vocalization.* As a level of emotional arousal, the number of distressful yelps and the duration of distress vocalization was recorded by one observer only while the other observer noted what else the pup was doing (random or nonspecific exploratory activities). There was a high correlation between distress vocalization and nonspecific exploratory activity.

After this 15-min testing period, the subject's approach to a passive observer in the arena was determined, and then approach and following response were assessed while the observer walked around the arena. The ability of the pup to negotiate a simple wire-mesh barrier placed between him and the observer was next used to assess detour behavior. Four trials were allowed, and if the subject was able to come around the barrier to the observer one end was blocked; if there was an end preference, the preferred side was blocked first. The time taken to pass around the barrier and the number of trials required were recorded. Finally, the social interaction of these differentially reared subjects was observed when they were placed together in the arena, which still contained the cloth and toy. After 5 min of observation, the experimenter entered the arena and observed the effects of the presence of a human on the group behavior of the pups. After these behavior observations, EEG recordings were taken on a Grass 6-channel recorder or on an Offner 8-channel type R dynograph, EEG recordings were also taken and evoked responses to visual and auditory stimuli of various frequencies were recorded on a computer of average transients (CAT 400 B) and monitored on an oscilloscope via a Grass 6-channel EEG recorder. Recordings were taken while the subjects were lying quietly awake and also while they were asleep in a darkened room; they were retained in a copper-gauze box lined with foam rubber. Selected subjects were then killed and several organs were dissected in the cold room at 37°F. The brain was dissected for histological examination, and the adrenal glands were dissected and prepared for epinephrine and norepinephrine analysis. Some tissues were also prepared for lipid analysis.

Appendix II

Litter I, comprising eight cubs, and Litter II, comprising four, were tested with a live adult rat placed in their home pen. Notes were taken as to which of the cubs were the first to respond by chasing, stalking, and seizing the prey and which were the most persistent during the 20 min observation period. A rating scale of 0–5 was used [0 = no reaction; 1 = orients; 2 = orients and approaches (stalks or chases); 3 = orients, approaches, and attempts to bite but shows ambivalence; 4 = orients, approaches, grabs, kills, but does not ingest; 5 = orients, approaches, grabs, kills, and eventually ingests]. This group test was repeated 24 hr later.

INDIVIDUAL PREY-KILLING

Each cub was tested in its nest box with a 5-week-old rat. Observations were cut off after 20 min, and the same rating scale described for the preceding test was used (see Fox, 1972a for further details of these and subsequent tests).

DOMINANCE RATING

Group and pair encounters were set up in the home cage at the normal feeding time using a caribou shank for the cubs to compete for. Tests of 20 min duration were run on alternate days for 8 days. As reported earlier (Fox, 1972a), this test is of value in identifying the most dominant and subordinate cubs. Middle-ranking cubs of Litter I, which showed greater proximity tolerance and "shifting" or relative dominance (in that one member of a pair appears to be dominant while in possession of food), were impossible to rank in a linear hierarchy. This test was most valuable in identifying the highest and lowest ranking cubs for later physiological tests. For this dominance test, and for the evaluation of fear reactions and exploratory behavior when presented with a novel stimulus, the same ratings as detailed earlier (Fox, 1972a) were used.

NOVEL STIMULUS

Using the same horizontal stimulus as developed earlier (a 30 cm × 2 m board with a 30-cm black square centered by a 10-cm white cross, Fox, 1972), the latency of response or emergence time from the nest box was recorded for each cub.

HEART RATE MEASURES

On the basis of the above tests, the two highest ranking and the lowest ranking cubs and one middle-ranking cub from Litter I were selected, and from Litter II (a more timid litter; see later) the highest and two lowest ranking cubs were used. Each subject was tested alone in a 4 × 4 × 3 ft arena; one observer remained behind an observation screen and recorded heart rates while the experimenter (E) went through a set procedure with each cub (see below). Heart rates were monitored biotelemetrically from a transmitter (E & M Instrument Company, transmitter type FM-1100-E2) taped to the subject's back and connected to skin electrodes secured with adhesive washers to a shaved area of the left thorax behind the left shoulder. The observer took four recordings of heart rates from the biotelemetry receiver each of 5 sec duration during the following procedures. (1) Initially a baseline heart rate was determined by taking sample EKG frequencies during the first 10 min in the arena. (2) E enters arena and sits for 20 sec diagonally opposite the cub, maintaining eye contact, but not moving. (3) E approaches cub and strokes head for 20 sec. (4) E backs off and resumes procedure (2). (5) E then picks cub up and holds it for 20 sec. (6) E leaves arena and after a 2 min interval, a baseline level is again taken. (7) E does not enter arena, but peeps around screen and makes direct eye contact with the subject for 20 sec. (8) A baseline rate is again taken after a 2-min interval. (9) Finally the heart rate change following a 5-sec duration bell (75 dB) placed outside the plywood arena next to the corner where the cub was lying is recorded.

STRESS STUDY

This study is based on the notion that confinement in a holding cage would be stressful and that cubs might react to such stress differently because of individual differences in emotional reactivity (which correlate with social rank). Blood samples were taken at arbitrarily selected intervals for plasma corticosteroid analysis with the anticipation that individual differences in the intensity and duration of the stress reaction might be found. Six cubs from the two litters, selected on the basis of high and low rank scores, were each placed in a 2 × 2½ × 2½ ft stainless steel cage that had been cleaned with Pinesol and lined with hay. The two cages were placed in a dimly lit and relatively soundproof room. Only two cubs were used for each run, 4 cc of blood being taken from the radial vein 5, 10, 30, and 60 min after confinement. Handling stress was kept to a minimum while these samples were being taken, this vein being easily accessible and the entire procedure taking no more than 30 sec. The hair over each foreleg had been shaved the previous day after termination of the EKG study in order to minimize additional handling stress. Twenty-four hours after these samples had been taken, four of the cubs (ScBl♂ and RcBr♂ of Litter I and ScBl♂ and BlUc♀ of Litter II) were injected with 20 units of ACTH at the time of confinement, and blood samples were again taken at the same intervals. The rationale behind this latter treatment was to attempt to elicit a maximal plasma corticosteroid response, and by comparing the data with the earlier samples, some indications as to response threshold and response magnitude as well as latency and duration might be disclosed.

Quantitatively collected blood plasma containing adrenocortical steroids was processed by solvent partitioning in (peroxide-free) ethyl acetate:methanol (99:1) on silica gel plates (Merck), elution of spots with 100% ethanol, and quantitations as described by Andrews (1968). Briefly, amounts of corticosterone from half of each eluate were estimated according to the fluorescence methods of Silber *et al.* (1958). The remaining half of the eluate was counted in a liquid scintillation spectrometer to determine loss of tracer-labeled steroids added to the samples prior to processing. Losses were corrected for in terms of label recovered from individual sample spots.

It should be emphasized that these cubs were not socialized to people: they had been raised with their mothers and were handled only once per week for approximately 15 min for making growth measurements. They were therefore essentially unsocialized but used to handling, during which they would manifest passive submission, occasional defensive aggression, and would attempt to escape whenever possible. All cubs showed the flight reaction when approached, and only the two highest ranking cubs would approach a passive person in the home cage to within a distance of 4–6 ft; none made contact or showed active submissive greeting. All cubs had received the same handling and no preferential treatment prior to the time of testing.

Appendix III

*From Linn, 1974. An Evaluation of Puppy Heart Rate and Plasma Cortisol Levels as Temperament Predictors in German Shepherd Dogs. Unpublished M.Sc. thesis, Washington University, St. Louis, Missouri.

SUBJECTS

One hundred and nine German shepherd dog puppies were tested in their 12th week of age. The puppies were from 21 different litters varying in size from 1 to 10 puppies in a litter. No bitch was used more than once. Six different males were used as studs.

SOCIALIZATION PROCEDURES

The puppies were removed from their mother in their 6th week. Usually the mother had begun the weaning from the 4th to 5th week. At 6 weeks of age the puppies were moved from the whelping area into either outdoor or indoor pens depending upon the weather and season of the year. At this time litters were dispersed as widely as possible. Usually one 10 × 12 ft pen contained up to 6 puppies from at least three different litters at 6 weeks of age. As they grew, the number in each pen was reduced until, at 6 months, there were two per pen. By the time they were 12 weeks of age, the age at which tests with which this study is most concerned were conducted, there were usually four per pen. At this age no attention was paid to the sex of the puppies when deciding where to place them. Each pen had

271

water and dry food *ad libidum*. In addition, a canned food and vitamin-mineral mix were fed to the puppies twice daily. On days with good weather and on which the puppy was not being tested, it was put out into a large wooded area with other puppies its size and allowed to explore and to play for approximately 5 hr. Water and dry food were available at all times in these areas. Up to 30 puppies were togther at one time. This procedure was followed until the puppies were 10 months old.

Starting on the 21st day of age the puppies were played with and handled in litters at least twice a week for at least 15 min each session. At weaning each pup was assigned to an individual handler. The puppies continued to be handled twice a week. They were walked on a leash and exposed to many different situations and environments such as automobiles, cats, rabbits, woods, and a lake. In addition, each pup was encouraged to "rough-house" with its handler, to play fetch, and to pull at a rag. Once each week each puppy was subjected to a formal evaluation session during which it underwent a battery of tests including "come" and "sit" commands, fetch, rag play, and a maze on the 8th, 9th, and 11th weeks.

At least once each week the puppies and their handlers chased a man (decoy) who was trailing a rag. The puppy was encouraged to bite the rag and to pull on it. There was generally a heightened sense of excitement in all of the puppies observing as well as the puppy doing the actual chasing. By 12 weeks of age most puppies were vigorously pursuing the decoy and biting the trailing rag. As the game progressed and the puppy matured, the rag was changed to a wrapping of cloth around the decoy's arm.

To summarize the puppies' experiences prior to novel experience testing, they were kennel raised, exposed to a large variety of situations, had been taught to be somewhat aggressive with humans, had never been scolded or treated harshly, and had been well socialized with humans and other puppies their size. Every effort had been made to prevent a stable status order from developing with cage mates. If there was an obvious dominance hierarchy developing within a pen, the pups in that pen were separated and put into the other pens.

EQUIPMENT

Novel Situation Apparatus

The novel situation apparatus was an 81-in. square room which had the floor marked into nine 26 in. squares separated by 3/4-inch black tape. The floor was painted gray and the walls were an off-white. Adjacent to the room, hereafter called the arena, and connected directly to it by a solid wooden guillotine door was a start-box 26 in. square and 18 in. high. The connecting door was 19 in. wide and 20 in. high. The start-box had a hinged top which was fastened down by a screen door-type latch. The arena was lighted by four 400-watt fluorescent tubes so as to light the center square with 145 foot candles. The start-box was not lighted. When the door to the start-box was opened and the lid raised, the lighting level in the box was 5 foot candles. The ceiling of the arena was 7½ feet from the floor. Observation was made through two one-way mirrors. One was in a corner by the start-box, and the other was directly opposite it. Both windows were 46 in. above the floor, high enough to prevent the puppies from seeing their reflection. Directly opposite the start-box entrance to the arena, a turntable was mounted with its base to the wall. The turntable was painted red and white. A table tennis ball and a tennis ball were suspended by a light chain from the top of the turntable in a manner such that, when revolving, the turntable hit the balls and caused them to bounce and the chains to emit a clanging sound. The turntable was set to 78 rpm. This apparatus is hereafter referred to as revolving stimulus. Below the turntable was a box constructed so that a cartridge-type tape recorder could be placed in it from outside the room. The tape used emitted a variety of noise levels ranging up to 80 dB (re .0002 dynes/cm^2) at the start-box arena interface. The frequencies ranged from 185 to 5600 Hz. The sound was constructed so as to vary high frequency, low frequency, and loudness in a random manner. In the middle of the wall to the right of the box entrance and to the left of the revolving stimulus and a foot above the floor was mounted an apparatus hereafter referred to as the rag stimulus. It consisted of a cylinder wrapped with rags. Around the rags was cotton rope tied so that loose ends dangled in the air in what was, hopefully, an enticing manner for the puppy.

The time spent in the box and arena was recorded on Cramer timers which timed to the nearest .01 min. Timers were mounted on the wall to be read outside each observation window.

Heart Rate Measuring Device

Heart rates were monitored with a Narco-Biosystems transmitter type FM-1100-E2. Recordings of the EKG were made by a Sanborn Poly Viso Recorder model 67-1200. The electrodes were moistened with electrode paste and secured with adhesive washers to the previous clipped area on either side of the thorax as in procedure.

Plasma Cortisol Equipment

Analysis of plasma cortisol was performed with the Schwartz/Mann Radioassay Kit using competitive protein binding. Calculation of precision gave a standard deviation in replicate dog plasma samples of $\pm$ 0.154 μg cortisol/100 ml plasma. Accuracy calculated as the standard deviation of percent recovery for 0.4 μg cortisol/100 ml plasma added as internal standard was found to be 93 ± 21%. The analysis was performed by the Army Environmental Hygiene Agency at Edgewood Arsenal, Maryland.

PROCEDURE

Not later than the day prior to testing, the hair was clipped over the left jugular vein and the right and left fifth and sixth intercostal spaces at the level of the elbow.

When the apparatus had been set up and tested the puppy was carried into the testing area and placed on a metal table. The puppy was positioned for bleeding from the jugular vein in such a manner as to place minimal stress on the puppy yet aid speed and ease of collection. The blood was collected in a heparinized 6-cc syringe with a 21-gauge 1½-in. needle. If the collection procedure was not an immediate success with a minimal amount of struggling by the puppy, that particular sampling attempt was immediately terminated. After collection of this baseline sample, the puppy was placed in the biotelemetry harness and the electrodes secured. The baseline heart rate was then obtained while the puppy was minimally restrained on the table top. The telemetered EKG was recorded for a 0.4-min period. Electrode connections and the telemetry apparatus were observed for proper functioning while recording the heart rate. The puppy was then immediately placed in the box and the lid secured. The puppy was left in the box for 1 min. Heart rate was recorded during the first and last 0.4-min periods while in the start-box. While the puppy was in the start-box the arena noise, turntable, and lights were turned on. After 1 min the guillotine door to the arena was raised and locked open. Four different EKG recordings were made during the 3 min the puppy was in or had access to the arena. The recordings were each 0.4 min long and were evenly spaced throughout the 3-min session. At the termination of the 3-min session, the puppy was called back and removed from the box. Two minutes following termination of the test the heart rate was again recorded as in the baseline. At the 5-min mark after termination of the arena testing, another heparinized blood sample was obtained from the jugular following the same procedure used in obtaining the baseline sample. The puppy had been restrained on the table top from removal from the box to the time of the 5-min blood collection. After that collection it was placed alone in a large pen adjacent to the testing area. Approximately 14 min following termination of the novel experience test, the puppy was retrieved from the pen and blood was collected at the 15-min mark.

Urination and defecation while undergoing the novel experience test was rare and subjectively appeared to be unrelated to the activity level or distress vocalizations of the puppies. Whenever elimination did occur, prior to the next puppy's session, the area was scrubbed and deodorized with a disinfectant solution. No sanitization procedures were utilized between sessions of most other puppies.

Obtained blood samples were centrifuged and the day's plasma collection sent to the laboratory for analysis. The EKG recordings were read for heart rate by counting the QRS spikes per unit of time.

TEMPERAMENT EVALUATION

Each puppy was evaluated continuously from birth. If a flaw of a physical nature such as hip dysplasia, elbow dysplasia, or cataracts or of a temperamental character such as shyness, fear of new situations, or dullness appeared, the dog was eliminated from further consideration as a breeder and, if the condition was severe, as a working dog. Actual evaluation after the 12th week of age was done by experienced personnel who were looking for a stable dog which could adapt logically to any situation it may come across. Adapt means perform according to the evaluator's ideals for a given situation. Generally the dog should be alert, curious, aggressive only when appropriate, have a high startle threshold, a high pain threshold, and show no fear in new situations. The evaluator gave pass/fail scores to all dogs at the time the decision was reached as to the future use of each dog. The score was on a scale of 3 to 0 with 3 and 2 being degrees of success and 1 and 0 being degrees of failure. Pass/fail and the 3 to 0 scores will appear throughout the results and discussion as appropriate. If, at 1 year of age, prior disposition had not been made of the dog, an evaluation score was also rendered on the basis of the dog's temperament at that time.

References

Ables, E. D. (1969). Home-range studies of red foxes (*Vulpes vulpes*). *J. Mamm.* **50**, 108–120.

Adams, D. B., Bacelli, G., and Mancia, G. (1968). Cardiovascular changes during preparation for fighting behavior in the cat. *Nature* **220**, 1239–1240.

Allee, W. C. (1949). *Principles of Animal Ecology*. W. B. Saunders Co., Philadelphia.

Andrews, R. V. (1968). Daily and seasonal variations in the adrenal metabolism of the brown lemming. *Physiol. Zool.* **41**, 86–94.

Astrup, C. (1968). Pavlovian concepts of abnormal behavior in man and animal. Pages 117–128 *in* M. W. Fox (Ed.), *Abnormal Behavior in Animals*. W. B. Saunders, Philadelphia.

Barnett, S. A. (1958). Experiments on "Neophobia" in wild and laboratory rats. *Brit. J. Psych.* **49**, 195–201.

Barnett, S. A. and Stoddart, R. C. (1969). Effects of breeding in captivity on conflict among wild rats. *J. Mamm.* **50**, 321–326.

Bateson, P. P. G. (1969). Imprinting and the development of preferences. Pages 109–124 *in* A. Ambrose (Ed.), *Stimulation in Early Infancy*. Academic Press, New York. 109–124.

Beach, F. A. and LeBoeuf, B. (1967). Coital behavior in dogs: I. Preferential mating in the bitch. *Anim. Behav.* **15**, 546–558.

Beck, A. M. (1971). The life and times of Shag, a feral dog in Baltimore. *Natural History* **80** (8), 58–65 (October).

Beck, A. M. (1973). *The Ecology of Stray Dogs: A Study of Free-ranging Urban Animals.* York Press, Baltimore. 98 pp.

Bekoff, M. (1972). An ethological study of the development of social interaction in the genus *Canis*—A dyadic analysis. Ph.D. Dissertation, Washington University, St. Louis, Missouri.

Belkin, D. A. (1968). Bradycardia in response to threat. *Amer. Zool.* **8,** 775 (Abstract).

Bennett, E. L., Diamond, M. C., Rosenzweig, M. R., and Krech, D. (1964). Chemical and anatomical plasticity of the brain. *Science* **146,** 610–619.

Berkson, G. (1968). Development of abnormal stereotyped behaviors. *Develop. Psychobiol.* **1,** 118–132.

Berry, R. J. (1969). The genetical implications of domestication in animals. Pages 207–217 *in* P. J. Ucko and G. W. Dimbleby (Eds.), *The Domestication and Exploitation of Animals.* Duckworth & Co., London.

Beylaev, D. K. and Trut, L. N. (1975). Some genetic and endocrine effects of selection for domestication in silver foxes. Pages 416–426 *in* M. W. Fox (Ed.), *The Wild Canids.* Van Nostrand Reinhold, New York.

Blauvelt, H. (1964). The physiological analysis of aggressive behavior (cited personal communication by D. E. Davis). Pages 53–74 *in* W. Etkin (Ed.), *Social Behavior and Organization Among Vertebrates.* University of Chicago Press, Chicago.

Bleicher, N. (1963). Physical and behavioral analysis of dog vocalizations. *Am. J. Vet. Res.* **24,** 415–427.

Bleicher, N. Personal communication. Downstate Medical Center, State University of New York, New York.

Blizard, D. A. (1971). Individual differences in autonomic responsivity in the adult rat. *Psychosom. Med.* **33,** 445–457.

Bowlby, J. (1953). Some pathological processes set in train by early mother–child separation. *J. Ment. Sci.* **99,** 265–272.

Bowlby, J. (1971). *Attachment.* Penguin, London.

Brunner, F. (1968). The application of behavior studies in small animal practice. Pages 398–449 in M. W. Fox (Ed.), *Abnormal Behavior in Animals.* W. B. Saunders, Philadelphia.

Burrows, R. (1968). *Wild Fox.* David & Charles, Newton Abbot, England.

Candland, D. K., Taylor, D. B., Dresdale, L., Leiphart, J. M., and Solow, S. P. (1969). Heart rate, aggression and dominance in the domestic chicken. *J. Comp. Physiol. Psychol.* **67**, 70–76.

Candland, D. K., Bryan, D. C., Nazar, B. L., Kopf, K. J. and Sendor, M. (1970). Squirrel monkey heart rate during formation of status orders. *J. Comp. Physiol. Psychol.* **70**, 417–423.

Cannon, W. B. (1956). "Voodoo" death. *Psychosom. Med.* **19**, 182–190.

Chance, M. R. A. (1962). An interpretation of some agonistic postures: the role of "cut-off" acts and postures. *Symp. Zool. Soc. Lond.* **8**, 71–89.

Chess, S. (1969). *Introduction to Child Psychiatry.* Grune & Stratton, New York.

Chiarelli, A. B. (1975). The chromosomes of the canidae. Pages 40–53 *in* M. W. Fox (Ed.), *The Wild Canids.* Van Nostrand Reinhold, New York.

Clutton-Brock, J. (1969). Carnivore remains from the excavations of the Jericho Tell. Pages 337–345 *in* P. J. Ucko and G. W. Dimbleby (Eds.), *The Domestication and Exploitation of Animals.* Duckworth & Co., London.

Corbett, L. and Newsome, A. (1975). Dingo society and its maintenance: a preliminary analysis. Pages 369–379 *in* M. W. Fox (Ed.), *The Wild Canids.* Van Nostrand Reinhold, New York.

Corson, S. A., O'Leary Corson, E., and Gwynne, P. H. (1975). Pet-facilitated psychotherapy. Pages 19–36 *in* R. S. Anderson (Ed.), *Pet Animals and Society.* Balliere Tindall, London.

Crisler, L. (1958). *Arctic Wild.* Harper Brothers, New York.

Darwin, C. (1873). *The Expression of the Emotions in Man and Animals.* Appleton, New York.

Darwin, C. (1875). *The Variation of Animals and Plants under Domestication,* 2nd Ed. John Murray, London.

Delius, J. D. (1967). Displacement activities and arousal. *Nature* **214**, 1259–1260.

Denenberg, V. H. (1964). Critical periods, stimulation input and emotional reactivity: A theory of infantile stimulation. *Psychol. Rev.* **71**, 351–355.

Denenberg, V. H. (1967). Stimulation in infancy, emotional reactivity and exploratory behavior. *In* D. H. Glass (Ed.), *Biology and Behavior: Neurophysiology and Emotion.* Rockefeller University Press, New York.

Denenberg, V. H. and Rosenberg, K. M. (1967). Nongenetic transmission of information. *Nature* **216**, 549.

Denenberg, V. H. and Whimbey, A. C. (1963). Infantile stimulation and animal husbandry: A methodological study. *J. Comp. Physiol. Psychol.* **56**, 877–878.

Eibl-Eibesfeldt, I. (1970). *Ethology: The Biology of Behavior.* Holt, Rinehart & Winston, New York.

Eisenberg, J. and Leyhausen, P. (1972). The phylogenesis of predatory behavior in mammals. *Z. Tierpsychol.* **30**, 59–93.

Ely, F. and Peterson, W. E. (1941). Factors involved in the ejection of milk. *J. Dairy Sci.* **24**, 211–233.

Enders, R. K. (1945). Induced changes in the breeding habits of foxes. *Sociometry* **8**, 53–55.

Engel, G. L. (1950). *Fainting.* Charles C. Thomas, Springfield, Illinois.

Epstein, H. (1971). *The Origin of the Domesticated Animals of Africa.* Africana Publishing Corp., London.

Ewer, R. F. (1973). *The Carnivores.* Cornell University Press, Ithaca, New York.

Fará and Catlett (1971). Cardiac response and social behavior in guinea pig. *Anim. Behav.* **19**, 514–523.

Fox, M. W. (1964). The ontogeny of behavior and neurologic responses of the dog. *Anim. Behav.* **12**, 301–310.

Fox, M. W. (1965). *Canine Behavior.* Charles C. Thomas, Springfield, Illinois.

Fox, M. W. (1966). *Canine Pediatrics.* Charles C. Thomas, Springfield, Illinois. 1966.

Fox, M. W. (Ed.) (1968a). *Abnormal Behavior in Animals.* W. B. Saunders, Philadelphia.

Fox, M. W. (1968b). Use of the dog in behavioral research. Pages 27–80 *in* W. I. Gay (Ed.), *Methods of Animal Experimentation, Vol III.* Academic Press, New York.

Fox, M. W. (1968c) The influence of domestication on the behavior of animals. *In* M. W. Fox (Ed.), *Abnormal Behavior in Animals.* W. B. Saunders, Philadelphia.

Fox, M. W. (1968d). Socialization, environmental factors and abnormal behavorial development in animals. Pages 332–355 *in* M. W. Fox (Ed.), *Abnormal Behavior in Animals.* W. B. Saunders, Philadelphia.

Fox, M. W. (1969a). Ontogeny of prey-killing behavior in *Canidae. Behavior* **35**, 259–272.

Fox, M. W. (1969b). The anatomy of aggression and its ritualization in Canidae: a developmental and comparative study. *Behaviour* **35**, 242–258.

Fox, M. W. (1970a). A comparative study of the development of facial expressions in canids: wolf, coyote and foxes. *Behaviour* **37**, 49–73.

Fox, M. W. (1970b). Neurobehavioral development and the genotype-environment interaction. *Quart. Rev. Biol.,* **45**, 131–147.

Fox, M. W. (1971a). *Integrative Development of Brain and Behavior in the Dog.* University of Chicago Press, Chicago.

Fox, M. W. (1971b). *Behaviour of Wolves, Dogs and Related Canids.* Jonathon Cape, London.

Fox, M. W. (1971c). Effects of rearing conditions on the behavior of laboratory animals. Pages 294–312 *in Nat. Acad. Sci. Defining the Laboratory Animal.* Washington, D.C.

Fox, M. W. (1971d). Ontogeny of socio-infantile and socio-sexual signals in canids. Z. *Tierpsychol.* **28**, 185–210.

Fox, M. W. (1972a). Socio-ecological implications of individual differences in wolf litters: a developmental and evolutionary perspective. *Behaviour* **41**, 298–313.

Fox, M. W. (1972b). *Understanding Your Dog.* Coward, McCann, New York. (Blond & Briggs, London. 1974.)

Fox, M. W. (1973). Social dynamics of three captive wolf packs. *Behaviour* **47**, 290–301.

Fox, M. W. (1974). *Understanding Your Cat.* Coward, McCann, New York. (Blond & Briggs, London. 1974.)

Fox, M. W. (1975a). Evolution of social behavior in canids. Pages 429–459 *in* M. W. Fox (Ed.), *The Wild Canids.* Van Nostrand Reinhold, New York.

Fox, M. W. (1975b) *Concepts in Ethology: Animal and Human Behavior.* University of Minnesota Press, Minnesota.

Fox, M. W. (1975c). Pet–owner relations. Pages 37–53 *in* R. S. Anderson (Ed.), *Pet Animals and Society.* Balliere Tindall, London.

Fox, M. W. (1976). The needs of people for pets. Pages 25–30 *in Canadian Symposium on Pets and Society,* Canad. Fed. Humane Societies, Toronto.

Fox, M. W. (1978). *In Search of Wildness & Whistling Jungle Dogs.* In preparation.

Fox, M. W. and Andrews, R. V. (1973). Physiological and biochemical correlates of individual differences in wolf litters. *Behaviour* **XLVI,** 129–140.

Fox, M. W. and Bekoff, M. (1975). The behaviour of dogs. Pages 370–409 *in* E. S. E. Hafez (Ed.), *The Behaviour of Domesticated Animals,* 3rd edition. Balliere Tindall, London.

Fox, M. W. and Clark, A. L. (1971). Ontogeny and temporal sequencing of agonistic behavior in the coyote, *(Canis latrans). Z. Tierpsychol.* **28,** 262–278.

Fox, M. W. and Cohen, J. A. (1977). Canid communication. *In* T. A. Sebeok (Ed.), *How Animals Communicate.* Indiana University Press, Bloomington, Indiana. In press.

Fox, M. W. and Spencer, J. (1969). Development of exploratory behavior in the dog: Age or experience dependent? *Develop. Psychobiol.* **2,** 68–74.

Fox, M. W. and Walls, S. (1973). Wild animals in captivity: Veterinarian's role and responsibility. *J. Zoo Anim. Med.* 4(3), 7–17.

Fox, M. W., Folk, G. E., and Folk, M. (1970). Physiological differences between alpha and subordinate wolves in a captive sibling pack. *Amer. Zool.* **10,** 487.

Fox, M. W., Lockwood, R., and Shideler, R. (1974). Introduction studies in captive wolf packs. *Z. Tierpsychol.* **35,** 39–48.

Funkenstein, D. H. (1955). The physiology of fear and anger. *Sci. Am.*

Gantt, W. H. (1944). *Experimental Basis for Neurotic Behavior.* Hoeber, New York.

Gantt, W. H., Newton, J. E. O., Royer, F. L., and Stephens, J. H. (1966). Effect of person. *Cond. Reflex* **1,** 18–35.

Gellhorn, E. (1968). Central nervous system tuning and its implications for neuropsychiatry. *J. Nerv. & Ment. Dis.* **147,** 148–162.

Gier, H. T. (1968). Coyotes in Kansas. *Kansas Agricultural Experiment Station Bulletin* **393,** 1–97.

Gier, H. T. (1975). Ecology and social behavior of the coyote. Pages 247–262 *in* M. W. Fox (Ed.), *The Wild Canids.* Van Nostrand Reinhold, New York.

Ginsburg, B. E. (1968). Genotypic factors in the ontogeny of behavior. *In* J. H. Masserman (Ed.), *Animal and Human.* Grune & Stratton, New York.

Graham, F. K. and Clifton, R. K. (1966). Heart-rate change as a component of the orienting response. *Psychol. Bull.* **65,** 305–320.

Gray, A. P. (1954). *Mammalian hybrids. A check-list with bibliography.* Technical Communication 10, Comm. Agriculture Bureaux, Farnham Royal, Bucks, England.

Guthrie, R. D. (1975). A hypothesis of density-adapted morphs among Northern canids. Pages 414–415 In M. W. Fox (Ed.), *The Wild Canids.* Van Nostrand Reinhold, New York.

Hale, E. B. (1962). Domestication and the evolution of behavior. *In* E. S. E. Hafez (Ed.), *The Behavior of Domestic Animals.* Williams and Wilkins Co., Baltimore.

Hale, E. B. (1969). Domestication and the evolution of behavior. Pages 22–42 *in* E. S. E. Hafez (Ed.), *The Behavior of Domestic Animals,* 2nd edition. Balliere Tindall, London.

Harlow, H. F. (1959). Love in infant monkeys. *Sci. Am.* **200,** 68–74.

Hediger. (1950). *Wild Animals in Captivity.* Butterworth, London.

Hediger, H. (1955). *Studies of the Physiology and Behaviour of Captive Animals in Zoos and Circuses.* Butterworth, London.

Henderson, N. (1970). Genetic influences on behavior of mice can be obscured by laboratory rearing. *J. Comp. Physiol. Psychol.* **72,** 505–511.

Hill, J. L. (1974). Peromyscus: Effect of early pairing on reproduction. *Science,* **186,** 1042–1044.

Hofer, M. A. (1970). Cardiac and respiratory function during sudden prolonged immobility in wild rodents. *Psychosom. Med.* **32,** 633–647.

Jensen, G. D. and Bobbitt, R. A. (1968). Implications of primate research for understanding infant development. *In* J. H. Masserman (Ed.), *Science and Psychoanalysis,* Vol. 12. Grune & Stratton, New York.

Joffe, J. M. (1969). *Prenatal Determinants of Behavior.* Pergamon Press, New York.

Joslin, P. W. B. (1966). Summer Activities of two timber wolf (*Canis lupus*) packs in Algonquin Park. Unpublished M.S. thesis, University of Toronto. 99 pp.

Keeler, C. (1970). Melanin, adrenalin, and the legacy of fear. *J. Heredity* **61,** 81–88.

Keeler, C. (1975). Genetics of behavior variations in color phases of the red fox. Pages 399–413 *in* M. W. Fox (Ed.), *The Wild Canids.* Van Nostrand Reinhold, New York.

Kennelly, J. J. and Roberts, J. D. (1969). Fertility of coyote-dog hybrids. *J. Mammal.* **50,** 830–831.

King, J. A. (1968). Species-specificity and early experience. Pages 42–64 *in* G. Newton and S. Levine (Eds.), *Early Experience and Behavior.* Charles C. Thomas, Springfield, Illinois.

Klinghammer, E. (1967). Factors influencing choice of mate in altricial birds. Pages 297–303 *in* H. W. Stevenson, E. H. Hess, and H. L. Rheingold (Eds.), *Early Behavior.* Academic Press, New York.

Kolenosky, G. B. (1971). Hybridization between a wolf and a coyote. *J. Mammal.* **52,** 446–449.

Kovach, J. K. and Kling, A. (1967). Mechanisms of neonate sucking behavior in the kitten. *Anim. Behav.* **15,** 91–101.

Kummer, H. (1971) *Primate Societies.* Aldine, Chicago.

Kuo, Z. Y. (1960). Studies on the basic factors in animal fighting. VII. Interspecies coexistence in mammals. *J. Genet. Psychol.* **97,** 211–225.

Kurtsin, I. T. (1968). Pavlov's concepts of experimental neurosis and abnormal behavior in animals. Pages 77–106 *in* M. W. Fox (Ed.), *Abnormal Behavior in Animals.* W. B. Saunders, Philadelphia.

Lacey, J. I. and Lacey, B. C. (1970). Some autonomic-central nervous system interrelationships. Pages 205–226 *in* P. Black (Ed.), *Physiological Correlates of Emotion.* Academic Press, New York.

Levine, S. and Mullins R. F. (1966). Hormonal influences on brain organization in infant rats. *Science* **152,** 1585–1592.

Leyhausen, P. *Verhaltensstudien an Katzen.* Paul Parey, Berlin and Hamburg, 1973.

Liddell, H. S. (1954). Conditioning and emotions. *Scientific American* **190,** 48–57.

Linn, J. (1974). An evaluation of puppy heart rate and plasma cortisol levels as temperament predictors in German shepherd dogs. Unpublished M.Sc. thesis, Washington University, St. Louis, Missouri.

Lockwood, R. (1976). An ethological analysis of social structure and affiliation in captive wolves. Unpublished Ph.D. thesis, Washington University, St. Louis, Missouri.

Long, E. M., Truex, R. C., Friedmann, K. R., Olson, A. K., and Phillips, S. J. (1958). Heart rate of the dog following autonomic denervation. *Anat. Rec.* **130,** 73–89.

Lorenz, K. (1966). *On Aggression.* Harcourt, Brace & World, New York.

Lorenz, K. (1968). *Evolution and Modification of Behavior.* University of Chicago Press, Chicago.

Lorenz, K. (1970). *Studies in Animal and Human Behavior*, Vol. II. Harvard University Press, Cambridge, Massachusetts.

Lorenz, K. (1975). Foreword in M. W. Fox (Ed.), *The Wild Canids*. Van Nostrand Reinhold, New York.

Lucas, E. A., Powell, E. W., and Murphree, O. D. (1974). Hippocampal theta in nervous pointer dogs. *Physiol. & Behav.* **12**, 609–613.

Lynch, J. J. (1970). Psychophysiology and development of social attachment. *J. Nerv. & Ment. Dis.* **151**, 231–244.

Lynch, J. J. and Gantt, W. H. (1968). The heart rate component of the social reflex in dogs: the conditional effect of petting and person. *Cond. Reflex* **3**, 69–80.

Macintosh, N. W. G. (1975). The origin of the dingo: An enigma. Pages 87–106 *in* M. W. Fox (Ed.), *The Wild Canids*. Van Nostrand, Reinhold, New York.

Marler, P. and Hamilton, W. J., III. (1966). *Mechanisms of Animal Behavior*. John Wiley & Sons, New York.

Mason, W. A. (1967). Motivational aspects of social responsiveness in young chimpanzees. Pages 12–34 *in* H. W. Stevenson, E. H. Hess and H. L. Rheingold (Eds.), *Early Behavior—Comparative and Developmental Approaches*. John Wiley & Sons, New York.

McBride, R. L., Klemm, W. R., and McGraw, C. P. (1969). Mechanisms of the immobility reflex ("animal hypnosis"). *Comm. Behav. Biol.* **3**, 33–59.

McNab, B. K. (1963). Bioenergetics and the determination of home range size. *Amer. Natur.* **97**, 133–140.

Mech, L. D. (1970). *The Wolf*. Natural History Press, New York.

Meggitt, M. J. (1965). The association between Australian aborigines and dingoes. Pages 7–26 *in* A. Leeds and A. P. Vayada (Eds.), *Man, Culture & Animals*. Publ. #78, AAAS, Washington D.C.

Meier, G. W. (1961). Infantile handling and development in Siamese kittens. *J. Comp. Physiol. Psychol.* **54**, 284–286.

Mengel, R. M. (1971). A study of dog-coyote hybrids and implications concerning hybridization in *Canis*. *J. of Mamm.* **52**, 316–336.

Menzel, R. and Menzel, R. (1948). Observations on the pariah dog. Pages 968–990 *in* B. Vesey-Fitzgerald (Ed.), *The Book of the Dog*. Borden Publishing, Toronto, Canada.

Meyer-Holzapfel, M. (1968). Abnormal behavior in zoo animals. Pages

476–503 *in* M. W. Fox (Ed.), *Abnormal Behavior in Animals*. W. B. Saunders, Philadelphia.

Morton, J. R. (1968). Effects of early experience on behavior. Pages 261–262 *in* M. W. Fox (Ed.), *Abnormal Behavior in Animals*. W. B. Saunders, Philadelphia.

Murphree, O. D., Dykman, R. A., and Peters, J. E. (1967). Genetically-determined abnormal behavior in dogs. *Cond. Reflex* **4**, 199–205.

Murphree, O. D., Peters, J. E., and Dykman, R. A. (1969). Behavioral comparisons of nervous, stable, and crossbred pointers at ages 2, 3, 6, 9 and 12 months. *Cond. Reflex* **4**, 20–23.

Nesbitt, W. H. (1975). Ecology of a feral dog pack on a wildlife refuge. *in* M. W. Fox (Ed.), *The Wild Canids*. Van Nostrand Reinhold, New York.

Newton, J. E. O. and Gantt, W. H. (1968). The history of a catatonic dog. *Cond. Reflex* **3**, 45–61.

Newton, J. E. O., Murphree, O. D., and Dykman, R. A. (1970). Sporadic transient atrioventricular block and slow heart rate in nervous pointer dogs. *Cond. Reflex* **5**, 75–89.

Nottebohm, F. (1970). Ontogeny of bird song. *Science* **167**, 950–956.

Obrist, P. A. (1968). Heart-rate and somatic motor coupling during classical aversive conditioning in humans. *J. Exp. Psychol.* **77**, 180–193.

Pavlov, I. P. (1928). (Gantt, W. H., transl.) *Lectures on Conditioned Reflexes*. International Publishers, New York.

Richter, C. P. (1954). The effects of domestication and selection on the behavior of the Norway rat. *J. Nat. Cancer Inst.* **15**, 727–738.

Richter, C. P. (1957). On the phenomenon of sudden death in animals and man. *Psychosom. Med.* **19**, 191–198.

Riesen, A. H. (1961). Stimulation as a requirement for growth and function in behavioral development. Pages 57–80 *in* D. W. Fiske and S. R. Maddi (Eds.), *Functions of Varied Experience*. Dorsey Press, Homewood, Illinois.

Rouget, Y. (1970). Personal communication. Department of Ethology, University of Rennes; film presented at International Ethology Conference.

Royer, F. L. and Gantt, W. H. (1961). The effect of different persons on the heart rate of dogs. Paper presented at Eastern Psychological Association, Philadelphia, March 1961.

Sackett, G. P. (1968). Abnormal behavior in laboratory-reared rhesus monkeys. *In* M. W. Fox (Ed.), *Abnormal Behavior in Animals.* W. B. Saunders, Philadelphia.

Sackett, G. P., Porter, M., and Holmes, H. (1965). Choice behavior in rhesus monkeys effect of stimulation during the first month of life. *Science* **147,** 305–306.

Salzen, E. A. and Cornell, J. M. (1968). Self-perception and species recognition in birds. *Behaviour* **30,** 44–65.

Schenkel, R. (1947). Expression studies of wolves. *Behaviour* **1,** 81–129.

Schneirla, T. C. (1959). An evolutionary and developmental theory of biphasic processes underlying approach and withdrawal. Pages 1–42 *in* M. R. Jones (Ed.), *Nebraska Symposium on Motivation.* Nebraska University Press, Lincoln, Nebraska.

Schneirla, T. C. (1965). Aspects of stimulation and organization in approach/withdrawal processes underlying vertebrate behavioral development. Pages 1–74. *in* D. S. Lehrman, R. A. Hinde, and E. Shaw (Eds.), *Advances in the Study of Animal Behavior,* Vol. I. Academic Press, New York.

Scholander, P. F., Irving, L., and Grinnell, S. W. (1942) Aerobic and anaerobic changes in seal muscles during diving. *J. Biol. Chem.* **142,** 431–440.

Schutz, F. (1965). Sexuelle Pragung bei Anatiden. *Z. Tierpsychol.* **22,** 50–103.

Scott, J. P. (1962). Critical periods in behavioral development. *Science* **138,** 949–958.

Scott, J. P. (1968a). Evolution and domestication of the dog. Pages 243–275 *in* T. Dobzhansky, M. K. Hecht and W. C. Steere (Eds.), *Evolutionary Biology,* Vol. II. Academic Press, New York.

Scott, J. P. (1968b) *Early Experience and the Organization of Behavior.* Brooks/Cole, Belmont, California.

Scott, J. P. (1973). *Animal Behavior.* Chicago: Chicago University Press, 1973.

Scott, J. P. and Fuller, J. L. (1965). *Genetics and Social Behavior of the Dog.* University of Chicago Press, Chicago.

Scott, M. D. and Causey, K. (1973). Ecology of feral dogs in Alabama. *J. Wildl. Mgmt.* **37,** 252–265.

Silber, R. H., Bush, R. D., and Oslapas, H. (1958). Practical procedure for estimation of corticosterone and hydrocortisone. *Clin. Chem.* **4,** 278–285.

Silver, H. and Silver, W. T. (1969). Growth and behavior of the coyote-like canid of Northern New England with observations on canid hybrids. *Wildlife Monographs* 17, 1–41.

Spitz, R. (1949). The role of ecological factors in emotional development. *Child Development* 20, 145–155.

Spitz, R. (1950). Anxiety in infancy: a study of its manifestations in the first year of life. *Int. J. Psychoanal.* 31, 138–143.

Staines, H. J. (1975). Distribution and taxonomy of the Canidae. Pages 3–26 *in* M. W. Fox (Ed.), *The Wild Canids*. Van Nostrand Reinhold, New York.

Tembrock, G. (1958). Lautenwiclung bein fuchs; sichter gemacht. *Umschau,* 58, 566.

Tembrock, G. (1960). Spezifische lautformen beim rotfuchs (vulpes vulpes) und ihre beziehunger zum verhalten. *Saugertierkundl. Mitt* 8, 150–154.

Tembrock., G. (1968). Land mammals. Pages 338–404 in T. A. Sebeok (Ed.), *Animal Communication*. Indiana Unversity Press, Bloomington, Indiana.

Theberge, J. B. and Falls, J. B. (1967). Howling as a means of communication in wolves. *Amer. Zool.* 7, 331–338.

Thomas, A., Chess, S., and Birch, H. G. (1970). The origin of personality. *Sci. Amer.* 223, 102–109.

Thompson, W. R. (1957). Influence of prenatal maternal anxiety on emotional reactivity in young rats. *Science* 125, 698–699.

Trumler, E. (1973) *Your Dog and You*. Seabury, New York. 1973.

Vandenberg, J. G. (1969). Male odor accelerates female sexual maturation in mice. *Endocrinology* 84, 658–660.

Van Lawick, H. and Van Lawick Goodall, J. (1971). *The Innocent Killers*. Houghton Mifflin, New York.

Vauk, G. (1953) Die abwandlung der Beutefanghandlung des hundes im zuge der domestikation. *Zool. Anzeiger (Deutsche Zoologische Geselscaft),* Supplement 17, 180–184.

Völgyesi, F. A. (1966). *Hypnosis of Man and Animal*. Ballière, London.

Wallace, R. K. and Benson, H. (1972) The physiology of meditation. *Sci. Amer.* 226, 84–90.

Weber, E. and Weber, E. H. (1845). Cited in Long *et al.* (1958).

Winter, P., Handley, P. Ploog, D. and Schott. (1973). Ontogeny of squirrel monkey calls under normal conditions and under acoustic isolation. *Behaviour XLVII,* Part 3–4, 230–239.

Wolf, S. (1964). The bradycardia of the dive reflex—a possible mechanism of sudden death. *Trans. Amer. Clin. Climat. Ass.,* 192–200.

Woolpy, J. H. Socialization of wolves, Pages 82–94 in J. H. Masserman (Ed.), *Animal & Human.* Grune & Stratton, New York.

Woolpy, J. H. and Ginsburg, B. E. (1967). Wolf socialization: a study of temperament in a wild social species. *Amer. Zool.* **7,** 357–364.

Zeuner, F. E. *A History of Domesticated Animals.* (1963). Harper & Row, New York.

Zimen, E. (1971). Wolfe und Konigspudel. *In* W. Wickler (Ed.), *Ethol. Studien.* Piper Verlag, Munchen.

Index